The Nature of Time

Edited by
RAYMOND FLOOD AND
MICHAEL LOCKWOOD

Basil Blackwell

First published 1986
First published in USA 1987
Reprinted 1988
First published in paperback 1988

Basil Blackwell Ltd
108 Cowley Road, Oxford OX4 1JF, UK

Basil Blackwell Inc.
432 Park Avenue South, Suite 1503
New York, NY 10016, USA

Library of Congress Cataloging in Publication Data

The Nature of time.
   Bibliography:p.
   Includes index.
   1. Time.   2. Space and time.   I.Flood, Raymond.
II. Lockwood, Michael.
BD638.N37   1986   115   86-13688
ISBN 0-631-14807-8
ISBN 0-631-16578-9 (pbk.)

British Library Cataloguing in Publication Data

The Nature of time.
   1. Time
   I. Flood, Raymond   II. Lockwood, Michael
   529   QB209
   ISBN 0-631-14807-8
   ISBN 0-631-16578-9 (Pbk)

Typeset in 10 on 12½ pt Century Text
by Pioneer Associates, Perthshire
Printed in Great Britain by
Billing & Sons Limited, Worcester.

Contents

The Contributors

Raymond Flood is Staff Tutor in Computing and Mathematics at the Department for External Studies, Oxford University. His main research area is the theory of probability; he is currently carrying out research for a book on cryptography.

Michael Lockwood is Staff Tutor in Philosophy at the Department for External Studies, Oxford University. His publications include *The Philosophical Imagination* (St Martin's Press, 1977) and *Moral Dilemmas in Modern Medicine* (Oxford University Press, 1985). He is currently writing a book on the mind—body problem.

Dennis Sciama FRS was for fifteen year a Senior Research Fellow of All Souls College, Oxford. He is now Professor of Astrophysics at the International School for Advanced Studies, Trieste. He is the author of *Modern Cosmology* (Cambridge University Press, 1971) and *The Unity of the Universe* (Faber, 1959).

W. H. Newton-Smith is Fellow and Tutor in Philosophy at Balliol College, Oxford, and is the author of *The Structure of Time* (Routledge and Kegan Paul, 1980) and *The Rationality of Science* (Routledge and Kegan Paul, 1981). He is currently involved in a project (partly funded by IBM) to introduce computers into the teaching of logic.

Roger Penrose FRS is Rouse Ball Professor of Mathematics at the University of Oxford. His main research interests are general relativity and its relation to quantum theory, and he has worked on black holes, twistor theory and quasi-crystallographic tilings; he is co-author, with W. Rindler, of the two-volume work *Spinors and Space—Time* (Cambridge University Press, 1984).

Michael Shallis is Staff Tutor in Physical Sciences, Department for External Studies, Oxford University. He received his doctorate in Astrophysics from Oxford University in 1978. Until 1970, when he began to study physics, he was a film director. He is the author of *On Time* (Penguin, 1983) and *The Silicon Idol* (Oxford University Press, 1984).

P. W. Atkins is University Lecturer in Physical Chemistry and Fellow of Lincoln College, Oxford. His recent books include *The Second Law* (Scientific American Library, 1984) and *Physical Chemistry*, 3rd edn (Oxford University Press and W. H. Freeman & Co., 1986).

Paul Davies is Professor of Theoretical Physics at the University of Newcastle and was formerly lecturer in applied mathematics at King's College, London. He is the author of *The Physics of Time Asymmetry* (Surrey University Press and the University of California Press, 1974) and several popular books on physics, including *Other Worlds* (Dent, 1980), *God and the New Physics* (Dent, 1983) and *Superforce* (Heinemann, 1984).

J. R. Lucas is a Fellow of Merton College, Oxford, and a tutor in philosophy. He is author of *A Treatise on Time and Space* (Methuen, 1973) and *Space, Time and Causality* (Oxford University Press, 1985), and is preparing, together with Dr P. E. Hodson, *Spacetime and Electromagnetism*, due to be published by Oxford University Press in 1987.

Michael Dummett has, since 1979, been Wykeham Professor of Logic at the University of Oxford. From 1962 to 1975 he was University Reader in Philosophy of Mathematics, and from 1975 to 1979, Senior Research Fellow, All Souls College. His recent publications include *The Interpretation of Frege's Philosophy* (Duckworth, 1981) and *Voting Procedures* (Oxford University Press, 1984).

1

Introduction

RAYMOND FLOOD AND
MICHAEL LOCKWOOD

This collection of essays is based on a series of lectures on the nature
of time given in Oxford in 1985 under the auspices of the Oxford
University Department for External Studies. The extraordinary
interest shown by the audiences, which, for some of the lectures,
numbered over 400, reflects the perennial fascination that time has
for humankind. Here, we have brought together some of the latest
thinking about time, both within philosophy and within modern
physical science. Modern physics and cosmology have, in fact, shed a
considerable amount of light on questions having to do with time
and space which would once have been thought of as the exclusive
province of the metaphysician. In this collection, time will be viewed
from the standpoints of Einstein's theory of relativity, modern
thermodynamics, quantum theory and contemporary cosmology, all
of which have some illumination to offer. It is also true, however,
that they in their turn have generated new conundrums for the
philosophers to worry about.

It may be helpful here to separate out some of the issues having to
do with the nature of time which are addressed in this volume. One
central issue is the basis of the so-called 'arrow of time'. Unlike space,
time seems to us to be inherently directional. Natural processes have
a natural temporal order. Physicists may tell us that it would violate
no natural law for water waves to converge on a pebble lying at the
bottom of a pond, and to expel it forcibly into the air, whence a
bystander might catch it in his hand. But this is something that we

never see in practice; whereas it is commonplace to see people throw stones into ponds, causing ripples to proceed outwards. Moreover, we have a special kind of knowledge about the past, reflected in our powers of recollection, and the wealth of traces that past events leave behind them. Also, we think of ourselves as being able to affect the future by our actions, in a way that we cannot affect the past. This, like our possession of memory, and the apparent absence of any corresponding powers of precognition, would normally be attributed to the fact that causation only works in one direction. Finally, we tend to think of the future as being in some sense *open*, by contrast to the past, which seems to us fixed, a *fait accompli*. Witness such proverbs as: 'It's no use crying over spilt milk.' The future, on the other hand — and this ties in with our conception of ourselves as possessing free will — is, in the words of an aphorism published some years ago in the *Reader's Digest*, thought of as 'not there waiting for us, but something we make as we go along'.

Two questions arise here. First, what, in philosophical or scientific terms, is the basis of this directionality? Secondly, to what extent is our ordinary thinking on these matters rationally sustainable, in the light of both modern scientific knowledge and philosophical analysis? The directionality of time represents a puzzle for physics, because the fundamental laws of nature, insofar as they are understood, appear (with one very minor exception) to be time-symmetric. Broadly speaking, if a process is physically possible, so is the same process run backwards. So why is it that, at the level of casual observation, the world is temporally so asymmetrical? It is customary here to appeal to the second law of thermodynamics, according to which *entropy* tends to increase over time. But this raises a number of further puzzles. First, what exactly is entropy, and why should it increase? These questions are addressed by Peter Atkins. Secondly, how can one get a temporally asymmetric principle such as this, if the fundamental underlying laws are time-symmetric? And, thirdly, how is one to reconcile the principle that entropy increases over time with the widespread view that the universe, at the big bang, was in a state of thermal equilibrium, which would normally be taken to mean maximum entropy? These latter two questions are examined by Roger Penrose.

A different approach to the problem of the direction of time is to be found in Paul Davies's article. Davies examines the direction of time from the viewpoint of quantum mechanics, relating it to the so-called collapse of the wave function. Whether it actually does collapse, however, and if so why, are, as Davies points out, controversial, puzzling and exciting questions. Davies ends by forging a connection between the wave particle duality and the traditional philosophical mind—body problem.

Both scientific and philosophical thinking about time have been revolutionized by the advent of Einstein's special and general theories of relativity. Einstein showed that space and time were intimately connected; indeed, the modern physicist thinks in terms, not so much of space and time, but of space—time, a unitary four-dimensional world, of which space and time are complementary but interdependent aspects. A consequence of the theory is that the rate at which time passes, as measured, say, by a perfectly adjusted clock, is crucially dependent on the velocity of the clock and the strength of the local gravitational field. These effects are explored in Dennis Sciama's article.

One traditional philosophical controversy has to do with the question of whether time and space are merely *relative* notions. Several philosophers have maintained that there is nothing to space and time over and above the spatial and temporal relations that hold between objects and events. Leibniz, for example, held this, as did Mach. Newton, by contrast, insisted that space and time existed independently of the objects and events that were in space and time: his view implied, contrary to Leibniz and Mach, that it made sense to ask whether an object was really moving, not merely relative to some other object, but absolutely, since either it was or it was not the case that it was at the same point of (absolute) space at different times. Seeing that Einstein's theory of relativity effectively does away with absolute motion, in this sense, it is often maintained that spatial and temporal relativism have won the day, vindicated by the march of physical science. Bill Newton-Smith, in a searching analysis of this issue, shows that such a conclusion would be premature. The very same philosophical issue of absolutism versus relativism which traditionally has arisen for space and time considered separately, arises also in the context of the theory of relativity, but now in relation to space—time.

We mentioned earlier the widely held view of the future as open, in contrast to the past, which is usually regarded as fixed. A philosophical articulation and defence of this traditional, indeed Aristotelian, view is given in John Lucas's article. It is a view which, in the light of Einstein's theory of relativity, may seem somewhat problematic. For one consequence of Einstein's theory is that simultaneity is relative. The collection of events simultaneous with our local 'now' defines the boundary between the past and the future. (The future is everything after now, the past everything earlier than now.) But Einstein showed that simultaneity is itself relative to one's choice of a frame of reference dependent on its state of motion. Observers moving relative to each other, but respectively thinking of themselves as at rest (which they are each equally entitled to do, according to Einstein), will, if they utter the words 'future' or 'past' as they pass each other, mean different things. For they will be slicing the space—time continuum at different angles. What is past for one observer may be future for the other, and vice versa. A proponent of the view that the future is open therefore, in the light of the theory of relativity, owes us an answer to the question: 'Future, relative to which frame of reference?' In short, the traditional view seems only to make sense on the assumption that there is some *preferred* frame. And to suppose this is to suppose something that, while it cannot be disproved, is contrary to the spirit of Einstein's theory.

Here, then, is a case of an aspect of our commonsense thinking about the future in relation to the past coming to seem doubtful, in the light of modern science. There is, however, another even more deeply entrenched aspect of our ordinary thinking about time which can be questioned on purely philosophical grounds. This is our ordinary, commonsense assumption that the past, in contrast to the future, cannot be influenced by what we now do. This assumption is challenged in a provocative paper by Michael Dummett. Dummett does not argue that we actually *can*, by our actions, influence the past. But he does show that the reasons usually offered for supposing that we cannot influence the past are philosophically unsound. Dummett's rather surprising philosophical conclusions might be thought to lend some weight to a recent claim by Wheeler, discussed in Davies's article, that certain quantum-mechanical phenomena actually involve backwards causation.

The authors so far referred to have discussed the nature of time either from a philosophical or a scientific perspective. Michael Shallis provides a timely (!) warning that our current scientific conception of time, quite as much as those of earlier civilizations, may reflect our own particular cultural biases. Impressed as we are with the progress of modern scientific inquiry, whose findings in relation to cosmology are explored in Shallis's article, we may take this conception to be closer to the 'objective' reality than it really is, or perhaps can be. For all that science or philosophy have to offer in the way of illumination, time remains profoundly enigmatic.

But it would be wrong to end on a negative note. The very fact that it is possible to bring the scientific method to bear on such fundamental, and to previous generations intractable, questions, is a measure of the extraordinary strides made by physical science in the course of the last 100 years. One might also conclude, looking at the articles contained here, that there is room for much fruitful collaboration between science and philosophy. On the one hand, physicists and cosmologists are continually finding that their researches lead them into areas where they are increasingly being forced to confront issues that are clearly philosophical in character. And on the other hand, science is producing findings that philosophers, in their own speculations, ignore at their peril.

Raymond Flood
Michael Lockwood

2

Time 'Paradoxes' in Relativity

DENNIS SCIAMA

INTRODUCTION

Time has always struck people as mysterious: mysterious, in fact, in a number of different ways. One thing that is mysterious about time is its directionality. What is it that underlies time's arrow? What, that is to say, is the source of the asymmetry between past and future, between earlier and later? Why, for example, can we remember the past but not the future? These are important and difficult questions. But they are not what I shall be talking about in this article. (The problem of the 'arrow of time' is, in fact, discussed by Roger Penrose in a later chapter of this book.) What I wish to talk about are the implications that Einstein's special and general theories of relativity have for our conception of time. What Einstein did, in effect, was to show that time is involved with the deep structure of physics, in a strange and entirely unexpected way.

The essential assumption underlying Einstein's special theory of relativity is that the velocity of light is independent of the motion of the observer. This seemed to be implied by Maxwell's equations, in which the speed of light appears as a constant; and was demonstrated empirically in the famous Michelson—Morley experiment. At first sight this seems a fantastic supposition. One is used to thinking, for example, that if one is travelling along in a car at 100 mph and fires off a bullet, in the direction one is travelling, at 700 mph, then relative to someone standing beside the road, the bullet will be travelling at 800 mph. So why shouldn't the same principle hold when, rather than firing bullets, one is firing photons, as indeed the car would be, if its headlights were on? In order to accommodate the assumption that light from a car's headlights will be travelling at precisely the same speed relative to the bystander and relative to the car, Einstein had to introduce a new law for the 'addition' — what is really meant is superimposition — of velocities. This has the consequence, not only that light from a car headlight will be measured as having the same velocity by the bystander as for the motorist, but also that the bullet, in our earlier example, will actually be travelling at very slightly less than 800 mph, relative to the ground. The discrepancy between what common sense would suppose and what Einstein's theory predicts gets progressively greater the larger the velocities involved, until one ends up with the consequence that 'adding' the velocity of light to (that is, superimposing it upon) any other velocity, up to and including the velocity of light itself, simply gives you the velocity of light. More generally, the superimposition of any two velocities less than that of light will produce a velocity which is itself less than the speed of light.

This may be difficult enough for the uninitiated to swallow. But worse still is to come. From the constancy of the speed of light (together with some other assumptions that no one would wish to quarrel with) Einstein was able to deduce, amongst other effects, that of *time dilation*. This is a phenomenon whereby the faster a clock moves, the slower it runs: the rate at which its hands move approaching zero as the clock's velocity approaches that of light, with respect to the people or instruments measuring the rate of the clock. This is a universal physical effect. It is not sensitive to structure, and applies to all processes whatsoever, including such biological processes as the rate of growth of hair, or ageing, and such physical processes as the decay of radioactive atoms.

Radioactive decay, in fact, provides a very good empirical demonstration of this effect. Cosmic rays, coming from outer space, generate μ mesons as they collide with particles in the upper atmosphere. Their rate of decay, or *half-life* (that is, the time you would have to wait before half the μ mesons in a beam of them would have decayed) acts as a clock. We know what the half-life of a μ meson at rest is; and a simple calculation shows that it is far less than the time it takes for a μ meson to get from the upper atmosphere to our detection apparatus on the surface of the earth. If the atmospheric μ mesons were decaying at the same rate as they do in the laboratory, when they are moving at only a tiny fraction of the speed of light, then we would expect to detect only about one-hundredth of the number we actually do. This is very strong evidence that for the μ mesons time has slowed down; and seeing that they are travelling at speeds very close to that of light, this is exactly what Einstein's theory predicts.

That is how *we* would interpret the effect, since in our frame of reference it is the μ mesons that are moving, and the earth which is at rest. But of course it is a key assumption in Einstein's theory that it is equally legitimate to take any unaccelerated frame of reference as the basis for one's description of physical processes: the same laws should apply. So it is reasonable to ask: how would a hypothetical observer actually sitting on a μ meson account for this effect? For him, after all, it is the earth, and we on it, that are moving at close to the speed of light and he that is stationary. Here we must appeal to another relativistic effect: the *contraction* of moving bodies. This is an effect whereby moving bodies become foreshortened in the direction of their motion. An observer relative to whom a rod is moving in the direction of its length will measure that rod as being shorter than an identically constructed rod relative to which he is stationary. (The older textbooks on relativity always used to say that such a rod would actually *look* shorter. But that is now known to be false. Relativistic optics in fact says that it will look rotated relative to the observer; but it is still true that when he takes appropriate account of the relativistic behaviour of light rays, he will *measure* the rod as being shorter.) An observer travelling with the μ meson will thus measure the distance between the upper atmosphere and the surface of the earth as being much less than we would think of it as being. For him, then, the reason why the μ meson has not decayed before it reaches the earth's surface is simply that it does not have very far to go.

We thus have two alternative ways of describing the same phenomenon, depending on which frame of reference we choose to base our description on. There is no contradiction between these two descriptions. The point is that, according to the theory, attributions of length, or of temporal separation between events, really only have an unambiguous sense when they are understood as relative to some choice of reference frame (though one could, if one wanted, define the true length of a rod to be its measured length in a frame of reference relative to which it was at rest). It should also be pointed out that, while observers that are moving relative to each other will tend to disagree about the time that has elapsed between two events, and the distance that separates them, there is something about which they will always agree: namely the so-called spatio-temporal *interval*. The reader will probably have heard time referred to as the 'fourth dimension'. Just as one can measure spatial distances, in three dimensions, by reference to three spatial coordinates — x, y and z — so in relativity one can measure space—time 'distances', known as intervals, in four dimensions, by reference to three spatial coordinates and one time coordinate. The results of such measurements turn out to be the same for all observers, regardless of their state of motion. They will disagree about the temporal and spatial separations taken individually, but when they come to put their different figures into the formula for calculating the interval, they will all arrive at the same answer. This is analogous to what happens, in three-dimensional geometry, if you have people working out the distance between two points using different systems of coordinates, rotated, say, with respect to each other. They will then disagree about the distances between the points, relative to their respective x, y and z axes; but they will all arrive at the same value for the overall spatial separation.

THE CLOCK 'PARADOX'

I turn now to the so-called *clock paradox*. Note that putting 'paradoxes' in inverted commas, as I have in the title of this article, is the writer's way of indicating that he intends to explain them in such a way as to show that they do not actually involve any inconsistency. And that is precisely my intention in regard to the clock 'paradox', which I shall now describe.

Suppose that a man leaves the earth in a spacecraft, flies to some distant point, and then returns to the earth. According to Einstein, as we have seen, the man ages less during this journey than the people who remain on earth. Likewise, a clock in the spacecraft registers a shorter time for the duration of the journey than does a clock on the earth. By way of illustration, let us suppose that the spacecraft's speed is one-seventh of a per cent less than the speed of light, and that when it returns people on earth are 20 years older than when it left. Then it follows from the special theory of relativity that the man himself will have aged only one year!

Now that space travel has become a reality, such a journey sounds more practicable than it did in 1905, when Einstein first devised the clock 'paradox'. It is no wonder, then, that there has been a revival of interest in Einstein's result (and certain similar effects arising in general relativity, to which I shall turn later). It can, in fact, be made even more startling: if the spacecraft were to fly yet closer to the speed of light, the man could return still a youth, to find his remote ancestors peopling the earth. Here is time travel indeed. Who would not rejoice at the possibility of sharing the human race's understanding of the universe 500 or 1000 years hence — or even knowing whether the human race has blown itself up by then? *But this is not yet the 'paradox'*.

To reach it, we must go one step further. As Einstein pointed out, and as I emphasized earlier, all motion is relative. This means that we can regard the man in the spacecraft as at rest throughout the journey, while the earth and its inhabitants shoot out and then return. But in this case we should expect the earthbound people to age less than the astronaut! Here then is the 'paradox': according to relativity a 'moving' person ages less than a 'stationary' one, but also according to relativity either person may be regarded as the 'moving' one. The apparent paradox arises because in Einstein's theory there is no *preferred* frame of reference, and hence no natural state of rest.

As a first step towards the resolution of this 'paradox' let us see with Einstein why the astronaut ages less when he is regarded as the one who moves out and returns. After this we shall examine what happens when the astronaut is considered to be stationary. To simplify the problem consider just two people, A and B. A is the stay-at-home, while B moves out in the spacecraft and then returns. Let us suppose that each of them carries a source of light which emits 50 waves per second in the other's direction. (This frequency does not actually correspond to visible light, but I want to keep the numbers simple.) By comparing what each of them sees of the other's light, we can, in fact, compare the amounts by which they age during the journey.

As the astronaut B moves out, he will see A's light Doppler shifted towards the *red*, that is, he will receive *fewer* than 50 waves per second. Let us suppose that he receives 49 waves per second (the actual number depends on his speed). When B turns around to begin the return journey, he will, *at that instant* see A's light change from being red-shifted to being blue-shifted, that is, as soon as he starts to move towards A he will receive more than 50 waves per second. If he returns at the same speed as he went out, he will in fact receive 51 waves per second. Thus B sees red for half his journey and blue for the other half. If each half of the journey takes, according to B, 100 seconds, then he receives altogether 10,000 waves (51 × 100 + 49 × 100).

Compare this with what the stay-at-home A sees of B's light. As B recedes, A will see the same red shift as B did on his outward journey, that is, A will receive 49 waves per second. But, and here is the crucial point, when B turns around and starts for home, *A will still see red light for a little while*. The reason for this is that at the moment of turnaround A is receiving light emitted by B *before* the turnaround, as the light takes time to reach A (see fig. 2.1). Indeed, when A first sees a blue shift, B will have performed more than half of his journey, and will be on his way home. This means that altogether A sees blue for a shorter time than he sees red, that is, waves arrive at the rate of 51 per second for a shorter time than they arrive at 49 per second.

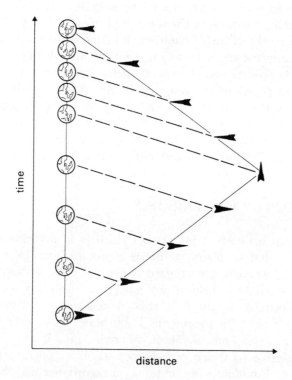

Fig. 2.1 *The Doppler effect in the light omitted by a rocket-ship which recedes and then returns. Low-frequency light is received by Earth for more than half the time*

How many waves does A receive altogether? If the whole journey lasted 200 seconds for A as well as for B, A would receive *fewer* than 10,000 waves, since the blue shift does not last long enough to compensate completely for the red shift. On the other hand, B emits 50 waves per second for 200 seconds, that is 10,000 waves, none of which can get lost. So A *must* receive these 10,000 waves by the time B reaches him, since B cannot overtake a wave moving with the speed of light. This is only possible if A receives waves for *longer* than 200 seconds. In other words, the journey lasts longer for A than it does for B: A has aged more than B. A detailed calculation, based on the formula for the Doppler shift, leads to the figures quoted at the beginning, namely, 20 years as against 1 year for a speed of a seventh of a per cent less than the speed of light.

This, in outline, is the argument used by Einstein to show that the moving person will age less than the stay-at-home. It is hardly surprising that it results in them ageing differently, since they are treated differently in the calculation. Whereas B is assumed to see the light change colour as soon as he changes his velocity, that is, at half-time, A sees a colour change *after* half-time. This seems reasonable in terms of the diagram of fig. 2.1. Nevertheless, the 'paradox' is still with us, since we can always describe the journey by saying that B remains at rest while A moves out and returns. In that case we might expect fig. 2.1 to apply with A and B interchanged. Why does it not? What is the difference between A and B?

RESOLUTION OF THE 'PARADOX'

The difference between A and B is that B is in a rocket-powered spacecraft — that is to say, he needs a powerful fuel to drive him away from A; whereas if we regard A as the person who moves, he still does not need the help of any such force. In other words, the difference between A and B is that B accelerates *relative to an inertial frame* when the rocket fuel is burning. An *inertial frame* is one which is itself unaccelerated: otherwise put, it is a frame of reference relative to which objects persist in a constant state of motion or rest if no forces act on them, in accordance with Newton's first law.

The crucial point about acceleration relative to an inertial frame is that it is something which admits of being detected just by inspection. If we're in a train, we can tell with our eyes closed that the train is slowing down or speeding up (although, unless we know whether we are facing the engine or not, we cannot without opening our eyes and looking out of the window tell *which* of these it is doing); we shall have the sensation of being pushed back into our seat, or else pulled forward away from it. Perhaps, our astronaut, B, in consequence of the acceleration he is experiencing, suffers from motion sickness (which is a misnomer for what is really *change of motion* sickness). Then that will be a pretty objective difference between A and B, and, by extension, between their two frames: one makes you sick, the other doesn't.

The situation is similar to that in Newton's famous *bucket experiment*. This was intended to demonstrate the existence of absolute motion, and hence of absolute space: that is, the existence of preferred inertial frames relative to which certain objects are 'really' unaccelerated. What Newton did was suspend a bucket from a hook by a long cord, twist it round several times, then fill the bucket with water and let it spin round, finally stopping the motion of the bucket to see what happened to the water. What he found, in a nutshell, was that the shape of the surface of the water was independent of its motion relative to the bucket. At the outset, the water was rotating relative to the bucket, but its surface was flat. The surface only began to become concave as the motion of the bucket was communicated to the water, which thus ceased to rotate relative to the rotating bucket, rotating instead with the bucket. Likewise, when the bucket was suddenly stopped, this made no immediate difference to the shape of the surface of the water, which remained concave for a time, becoming flat only as friction with the side of the bucket progressively slowed its rotation.

We can sum up Newton's interpretation of his experiment by saying that absolute rotation has nothing to do with the relative rotations which are directly observed, and that, nevertheless, we *can* determine experimentally the amount of absolute rotation possessed by a body. All we have to do is measure the curvature of a water surface rotating with the body. And rotation, of course, is just a special case of acceleration. (Bear in mind that acceleration means change of *velocity*, and velocity is defined as motion in a given direction. When something is rotating, its parts are moving at a constant *speed*, but since they are moving in a circle they are constantly changing their direction of motion, and hence their velocity.)

Newton's reasoning is flawed, however. His experiment does not demonstrate that there is absolute acceleration. For, as Bishop Berkeley pointed out, there is another way of interpreting his result. The reason that the surface of the water rises may not be that it is rotating absolutely, but that it is rotating relative to the 'fixed stars' that is, to the bulk of the rest of the universe. This idea was taken up by Ernst Mach in 1872. According to Mach's principle, all acceleration is relative acceleration. It would follow from Mach's principle that inertial frames are distinguished not by being absolutely unaccelerated, but rather by being unaccelerated relative to some suitably defined average of all the matter in the universe.

If we applied this idea to the present problem, we could say that the difference between our two people is that B has accelerated relative to distant matter while our stay-at-home A has not. In an otherwise completely empty universe, the situation as between A and B would be symmetrical — if Mach's principle is correct — and either could be considered to be the accelerated one. In that case there should be no difference in their ageing. But in the presence of stars and galaxies, Mach's principle entitles us to regard B as accelerated and A not.

These considerations dispose of the clock 'paradox'. Whether or not one accepts Mach's explanation for the difference between inertial and non-inertial frames, there clearly is a difference. The clock 'paradox' only seems paradoxical so long as one neglects that difference, which is, as we have seen, absolutely crucial to explaining why B has aged less than A. But not, it must be emphasized, because acceleration *itself* has any effect on ageing or the rate at which clocks run; rather, because it places the astronaut B and the stay-at-home A in an asymmetrical relationship to each other.

That, however, raises a further question: can we be sure that acceleration does not, in its own right, give rise to a time-dilation effect? Let me stress that I am not talking, here, about any purely *mechanical* or structure-sensitive effect that acceleration might have on a clock; let's assume that we're dealing with a *robust* clock and only moderate forces. According to Einstein, what a robust clock, if accurate, does is measure elapsed *proper time* along its own *world-line*: its path, that is, through space—time. The claim is that if we take any two events on the clock's world-line, then the time that, according to the clock, has elapsed between the two events will be proportional to the spatio-temporal length of that segment of the clock's world-line that is bounded by these points. (If the spatio-temporal length is measured in spatial units, then dividing it by the velocity of light will, by definition, give you the elapsed proper time.) From this point of view, the fact that, for A, more time has passed than it has for B is wholly attributable to the fact that A's world-line, between his departure and his return to earth, is, in spatio-temporal terms, shorter than B's. And this, in turn, is a consequence of the geometry of space—time.

It is important to emphasize, however, that it is an empirical *assumption* that what a robust clock measures is proper time. It is an assumption, moreover, that has sometimes been questioned by physicists. In 1918, Weyl put forward an ingenious generalization of Einstein's general theory of relativity, designed to accommodate electromagnetism, which predicted that clocks would be affected by the electromagnetic field along the path of an object's world-line. This theory is no longer accepted. Einstein pointed out that, were the theory correct, we should expect to find, as we do not in fact find, that the spectral lines of atoms on the sun were different from those of the corresponding atoms on earth, even when appropriate account had been taken of the gravitational red shift and Doppler shift. Nevertheless, Weyl's theory has a certain historical importance as the first example of a *gauge theory* in physics. Contemporary theories of the elementary particles are all, in the mathematical sense, gauge theories. This is by the way. But now, might it be that clocks really are affected by acceleration, in the way that Weyl theorized, mistakenly, that they were affected by the electromagnetic field?

It is perfectly possible to test this experimentally. Once again, we can use radioactive decay as our 'clock', seeing what happens when radioactive particles with a known half-life are speeded up in a particle accelerator. Now particles in an accelerator travel in a circular path. As I remarked earlier, acceleration means change in velocity, and velocity is speed in a given *direction*. Particles travelling in a circle are constantly changing their direction; consequently they are undergoing acceleration, even when they have settled down to a constant speed. When one observes such particles — the experiment has been done using μ mesons — one finds that the time dilation is just what one would expect, given their velocity. There is no additional effect that could be attributed to the acceleration of the particles. This negative result is corroborated by the so-called *Mössbauer effect*, whereby certain bound atomic nuclei in solids can be made to fluoresce in the gamma ray part of the electromagnetic spectrum. These nuclei are constantly vibrating, changing their state of motion, and are hence subject to a very considerable acceleration. Scientists have made careful measurements of the shift in frequency of the emitted radiation (using atoms of the isotope ^{57}Fe, of iron) and have found a *velocity-dependent* frequency shift that matches the predictions of special relativity. But the acceleration, *per se*, was associated with no observed time-dilation effect; if it existed at all it would have had to account for less than one part in 10^{13} of the total frequency shift.

GENERAL RELATIVITY

That said, velocity is not the only thing that gives rise to a time-dilation effect. According to Einstein's *general theory* of relativity, gravitation also causes clocks to slow down. Observers in a less intense gravitational potential find that clocks in a more intense gravitational potential run slow relative to their own clocks. This effect can be demonstrated in various ways. One is by looking at the radiation emitted by atoms at the surface of the sun, where the gravity is much greater than at the surface of the earth. What one finds is that the radiation is shifted towards the red end of the spectrum as compared to the radiation emitted by similar atoms on the earth. This is known as the *gravitational red shift*. The effect can even be observed in the laboratory, by comparing frequencies of identical sources at the ceiling and the floor. Even though the difference in the strength of the gravitational potential is minuscule, the resultant red shift is nevertheless sufficiently great to be detectable by modern instruments. (The gravitational red shift was first demonstrated in the laboratory in 1965, using the Mössbauer effect, with apparatus mounted at the top and bottom of a 22 metre tower at Harvard.)

No discussion of these matters nowadays would be complete without some mention of *black holes*. These are bodies that are so compact for their mass that light cannot leave their surface against the intense gravitational field. What, then, about the gravitational red shift of a black hole? Well the general theory of relativity predicts that it should become infinite as one approaches the point at which light is no longer able to escape: the so-called *event horizon*. At this point clocks, from the standpoint of a distant observer, have stopped completely: there is an infinite time-dilation effect.

This has very strange consequences. It means that if an astronaut takes off from earth and flies towards a black hole, then, from the standpoint of an earthbound observer, he takes an infinite time to get there: though he gets ever closer to the event horizon, he never actually reaches it. And if we could somehow observe the astronaut's watch, through a very powerful telescope, we would see it go slower and slower, until eventually it was going so slowly that it would appear to have stopped completely. The astronaut would seem to hover at the event horizon, frozen in time. (At least, that is how it would seem were we able to see him; the light coming from his spacecraft would in fact have become so red-shifted by then that he would no longer be visible at all.)

All this is as measured by our clocks. That is to say, if we take an earthbound clock to define our time coordinate, then the corresponding *coordinate time* it takes for the astronaut to reach the event horizon is infinite. But coordinate time must be contrasted with the astronaut's *proper time*: time as registered by the astronaut's own clock, measuring off the spatio-temporal distance along his own world-line. From the astronaut's own point of view, it takes only a finite time to reach the event horizon and cross over into the interior, whence he can never return, since to do so he would have to exceed the local light velocity. His inevitable fate is then, in a very short proper time, to be torn apart by tidal forces as he is inexorably drawn towards the singularity at the centre of the black hole — a point at which the density of matter becomes, in theory, infinite.

I must stress that this slowing down of all physical processes aboard the astronaut's spacecraft is a perfectly genuine effect, and not merely an optical illusion engendered by the influence of an intense gravitational field on the behaviour of light rays. Suppose the astronaut were to have second thoughts shortly before crossing the event horizon, and had sufficiently powerful rockets to escape from the gravitational pull of the black hole and return to earth. Then he would be found to have aged less than a twin brother who had stayed at home, who might well, by then, have died of old age.

A FURTHER EFFECT OF ACCELERATION

I want to end with a brief discussion of a phenomenon that has been known about for less than ten years, but has a considerable bearing on the previous discussion. In 1976, Unruh showed that an object accelerating in a quantum vacuum would find itself in a 'heat bath': it would be irradiated with electromagnetic radiation with a so-called *thermal* or *black-body* spectrum — similar, that is to say, to that radiated by an object, at the temperature in question, which was in perfect equilibrium with its surroundings, absorbing as much radiation as it emitted. The temperature of this radiation to which an accelerated body would be subject can be shown to be proportional to the acceleration and to be independent of the velocity.

Clearly, this provides us with another asymmetry between our stay-at-home A and astronaut B in the clock 'paradox'. But how are we to understand this effect? Well it has to do with the so-called zero-point energy, or quantum fluctuations of the vacuum. According to quantum theory, the vacuum is constantly emitting and absorbing radiation in a random fashion. And so also is any body passing through it. But if we take a detector that is moving through the vacuum at a constant velocity, it will not register any net radiation; for the vacuum fluctuations are effectively cancelled out or compensated for by the random fluctuations, or 'noise', within the detector itself. When the detector is accelerating, however, this balance is upset — the zero-point fluctuations of the ambient electromagnetic field are no longer fully compensated for by the noise in the detector, and it begins to register a signal.

It is impossible, obviously, to provide a complete explanation of why this is so, without getting into technicalities that would be intelligible only to those familiar with quantum-field theory. The crux of the matter — the crucial fact that makes the calculations come out the way they do — is, however, easy to explain. As is shown in fig. 2.2, the world-line of an object that is uniformly accelerated — subject, that is, to a constant acceleration — takes the form of a curve: specifically, a *hyperbola*. If we take any two points on this curve, then the elapsed time measured along the curve is different from that measured along a *geodesic* joining them, represented in the diagram as a straight line. (The world-line of an unaccelerated body is invariably a geodesic.) It is precisely this difference in elapsed time between any two such points, as measured along the detector's world-line, and as measured along a geodesical or unaccelerated path, that generates the heat-bath effect, by creating an imbalance between the vacuum fluctuations in the field and those in the detector.

As I say, this effect creates a further asymmetry between the stay-at-home A and the astronaut, B; B, as he turns around to come home, will experience a rise in temperature, however well insulated the cabin of his craft is from the rocket engines. As a matter of fact,

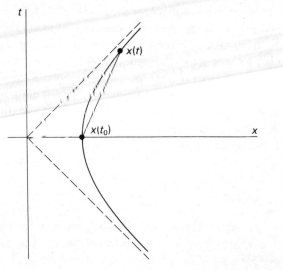

Fig. 2.2. *The space—time path of the uniformly accelerating detector is a hyperbolic curve. The elapsed time between two representative points on the path is different if measured along the path or along the geodesic joining them. (The dotted line running from the lower left to the upper right is an event horizon for the detector: signals originating to the left of this line can never reach the detector, as long as it continues in a state of uniform acceleration.)*

there is a deep and instructive connection between this phenomenon and another phenomenon that will be familiar to some readers. I have in mind the *Hawking radiation* now known to be emitted by black holes. Black holes are not, in fact, completely black. They emit black-body spectrum radiation, and their event horizons have, in consequence, a non-zero temperature. This effect is too small to be detectable in a black hole of, say, one solar mass. But it becomes explosive when the black hole is sufficiently small. Black holes decay as they radiate (unless they are replenished by infalling matter or radiation). As they decay the intensity of the emitted radiation inexorably rises — the black holes get hotter — with the result that they eventually go out with a considerable bang, in an explosive burst of radiation.

Now the reason why black holes radiate is, from a certain point of view, essentially the same as the reason why accelerating bodies are subject to radiation. (It ties in with Einstein's observation that an observer subject to a constant acceleration would be unable to distinguish this from a gravitational force.) An observer who is undergoing constant acceleration will find that he has an event horizon, just as a black hole has an event horizon: that is, a point beyond which certain regions are unable to communicate with the observer. Once again, to explain precisely why this is so is beyond the scope of this paper. Suffice it to say that these parallels illustrate, once again, the extraordinary elegance and essential underlying unity of Einstein's theory.

DISCUSSION

Q. You say that acceleration does not give rise to a time-dilation effect, but that gravity does. This puzzles me. I'd always understood that, according to Einstein's *equivalence principle*, it is impossible for an observer to distinguish between being in a state of acceleration and being subject to gravitation. Doesn't it follow from that principle that if gravity gives rise to a time-dilation effect, acceleration ought also to do so?

A. Einstein claimed that it is impossible for an observer to distinguish between being accelerated and being subject to a gravitational 'pull'. If an astronaut is in a soundproofed spacecraft without windows, there is no experiment he can perform which will tell him whether he is sitting on the launching pad and subject to gravitation, or in deep space with his rockets blasting away. However, the time-dilation effect has nothing to do with the effects of a gravitational pull *at one place*. The gravitational time-dilation effect arises from *differences* in gravitational potential between one place and another: the observer in the less intense gravitational potential measures clocks in a more intense potential as going slow relative to his own clocks. When I said that acceleration does not give rise to a time-dilation effect, I meant a *local* effect, that is, as measured by a nearby but unaccelerated observer. Einstein's time-dilation due to velocity is, in this sense, a local effect.

Note: This was written up by M. L. with the aid of several sets of notes kindly supplied by people who had attended the lecture. It incorporates some extended passages, including most of one chapter, from Dr Sciama's book, *The Unity of the Universe*, and has been revised by Dr Sciama himself.

FURTHER READING

Sciama, D. W., *The Unity of the Universe* (Faber and Faber, London, 1959), especially chapters 7—11. This book is unfortunately out of print, but should be available through libraries.

A technical article which covers in far greater detail the 'radiation bath' phenomenon discussed at the end of this chapter is: Sciama, D. W., Candelas, P. and Deutsch, D., 'Quantum field theory, horizons and thermo-dynamics', *Advances in Physics*, 30 (1981), pp. 327—66. Though written for the specialist, this article — particularly sections 1 and 2 — should prove accessible to the general reader.

Einstein, A., *Relativity, the Special and General Theory: A Popular Exposition* (Methuen, London, 1920).

Marder, L., *Time and the Space Traveller* (Allen and Unwin, London, 1971).

3

Space, Time and Space—Time: A Philosopher's View

W. H. NEWTON-SMITH

To ask a philosopher to provide an account of something like space, time or space—time is perhaps a little unfair. For philosophers are better at asking difficult questions than in agreeing on the correctness of a particular philosophical account. Notoriously they are prone to agonize about questions of the form — what is x? — where 'x' is replaced by, say, 'Truth', 'Beauty', 'Goodness'; or, in the case at hand by 'Space', 'Time' or 'Space—time'. These questions are curious. For in one sense we already know the answers. You have no problems in understanding me if I say that the *next time* a political figure is offered a chance to be proposed for an honorary degree, she will decline. Or to say that *every time* Professor Snerd lectures there is a lot of empty space in the lecture hall provokes no failure to understand what was meant. Our problem is that we cannot say what it is that we understand by words that we obviously do in one sense understand perfectly well. In this regard our predicament is rather like that of a bicycle rider who can be in a position of skill but completely unable to say what it is he actually does to keep the bicycle upright. Indeed, focussing too much on this question while exercising the skill may interfere with the exercise of that skill.

To some this analogy will rather count against the philosophical enterprise. For one does not need an account of how it is that one rides a bicycle in order to get on with cycling. And that suggests that we simply get on with developing our scientific theories of space, time and space—time without bothering about saying what we mean by 'space', 'time' and 'space—time'. However, that is but a weakness in the analogy. For one conclusion which will emerge from this paper is that the resolution of certain debates in physics requires the resolution of certain philosophical issues that arise when we pursue these questions. Among those who thought that such philosophical questions needed answering was Isaac Barrow, whose thinking on these topics greatly influenced his student Isaac Newton.

But because Mathematicians frequently make use of Time, they ought to have a distinct Idea of the meaning of that Word, otherwise they are Quacks. My Auditors may therefore, on this Occasion, very justly require an Answer from me, which I shall now give, and that in the plainest and least ambiguous Expressions, avoiding as much as possible all trifling and empty words. [From Barrow; see Further Reading.]

My theme is that Barrow is correct. Unless we have a philosophically adequate understanding of what we mean by these concepts we will be but quacks and will be prone to make mistakes in our physics as quacks make mistakes in medicine. Equally our philosophical understanding will be deficient if it does not take account of physical theory.

We can in fact learn something about space and time by first considering why it is difficult to say what we understand by 'space' and 'time'. And this will emerge if we compare the questions at hand with more tractable questions. Suppose, for example, I was asked what a cuttlefish is. Given a handy aquarium one natural way to proceed is to use ostension; that is, to provide an example of a cuttlefish. Such a procedure is not open to us in the case of space and time. We can neither point to space nor time nor to the points that make up space and time. Such items as locations in space or time are not items of our experience. If we try to think of a moment of time, for instance, the best we can do is to think of an event which happened at some moment. Like Hume's soul, the moment itself entirely eludes our grasp. We can express the problem here by saying that space, spatial items like locations, time and temporal items like moments or instants are *abstract*. For anyone with an empiricist orientation such items are bound to seem problematic for they cannot be displayed or studied in any direct fashion.

Another procedure which we might seek to use in the face of such questions is to give a verbal definition. If I am asked what we mean by a 'zemindar' I might well say that a zemindar was a revenue officer of the Mogul empire. This suggests that we might turn to the *Concise Oxford Dictionary* for enlightenment. If we look up the entry for 'time' it says: continued duration. But if we then turn to the entry for 'duration' we find it written that it is a part of time! Such a tight little circle of definitions is hardly enlightening. Aristotle's definition of 'time' as 'the number of motion' is equally circular, motion being defined itself by reference to change of location in space with regard to time. Not only have such neat verbal formulations tended sooner or later to be circular, they have also tended to explain the obscure notions of space and time in terms of equally if not more obscure notions. Other attempts to define time in terms of causation or consciousness manage to be both circular and obscure. This problem arises from the *promiscuous* character of the notions of space and time. These concepts are so intimately connected with such a wide range of fundamental ideas that the prospects of finding some unproblematic terms with which to define them are dim. If we turn our attention to the concept of space—time the prospects of gaining any easy insight are even more remote. For while the concepts of space and time are in some sense familiar, the concept of space—time, as we shall see, depends on highly technical theory and involves strange and counter-intuitive elements. What then is to be done? Augustine in a frequently quoted passage said that he understood what time was so long as no one asked him. But if asked what he understood by time he did not know. This reduced him to praying for enlightenment. In a more secular age the way forward is to consider what I will call theories of space and time. By theories I mean general accounts of space and time that are meant to cover *some* aspects of these notions through an account of the relation between space and time on the one hand and the physical world on the other.

Caroline, Princess of Wales, provided an important introduction to this approach when in the early eighteenth century she acted as an intermediary between Leibniz and the Reverend Samuel Clarke. Leibniz wrote to her lamenting the decay in England of natural religion, a phenomenon for which he held Newton responsible. Caroline passed the letters to Clarke who answered on Newton's behalf. In the ensuing exchange of letters a major controversy about space and time emerged which is as open and lively today as it was in the days when Princesses of Wales took an interest in such matters. And this provides a convenient starting point for us. For the debate takes place in the context of classical physics and enables us to get a sense of the issue without bringing in the more demanding contemporary physical theories. A brief historical reconstruction of that debate will display the importance of an interplay of philosophy and physics in discussions of space and time. With that in the background we will be able to see that the same issue arises with regard to space—time and that the contemporary debate similarly involves both philosophical and physical issues.

To characterize the Leibnizian position we need to distinguish between *things in time* such as events and *temporal items* such as moments and instants. In the case of space we similarly distinguish between things in space such as bodies and the spatial items such as locations. Initially, we can think of time as a system of temporal items such as instants and space as a system of spatial items such as locations. Leibniz's fundamental claim was that time is nothing over and above an ordered system of events and that space was nothing over and above a system of bodies. We can describe Leibniz's approach as *reductionistic*. Focussing on time for the moment, Leibniz thought that all talk about time or temporal items could be translated without loss of meaning into talk about events and various relations between events.

An analogy will be helpful in bringing out the content of a reductionist approach. It may well be true to say that the average reader of this book has 2.2 children. If I search through the readers eager to meet the average one with his or her 2.2 children I will be disappointed. And if I cling to the view that there is such a person that person will come to seem most mysterious being invisible and intangible. My mistake, reminiscent of Lewis Carroll, arose from taking 'the average reader' on a par with 'the youngest reader'. In fact there is no such person as the average reader over and above the actual readers of this paper. The original truth is to be construed as saying that the total number of children divided by the total number of readers yields the result 2.2.

The Leibnizian approach is to treat such assertions as 'Time had a beginning' or 'The *moment* I began to lecture you yawned' as requiring a translation into assertions which make reference only to events and to temporal relations between events. The former would be construed as the claim that there was an event with no events before it and the latter would be construed as saying that the set of all events simultaneous with my beginning to lecture contained your yawning. Thus, for Leibniz there is nothing to time or temporal items over and above the events and processes that we would normally say occurred in time. And, similarly, all assertions about space or about spatial locations would be translated into assertions about bodies and spatial relations between bodies. Talk of space, spatial items, time and temporal items is a mere *façon de parler* which it is convenient to use in the same way that the phrase 'the average reader' serves to give economy of expression.

Since all talk of space and time is to be construed as talk about events and objects, the investigation of space and time becomes an empirical matter. Whether time had a beginning is something to be settled by reference to the cosmological question as to whether there was a first event in the history of the universe. And the question of the structure of space, whether it is Euclidean or not, will require a physical investigation into the contents of space. Such an approach to space and time is attractive for at least two reasons. First, it is ontologically parsimonious. We do not need to admit the existence of space and time as items over and above events and bodies. Secondly, it is demystifying. We do not have instants of time and points of space as mysterious items not given in experience. Events and bodies are types of items of which we can have experience and we can restrict attention to them and to items constructed out of them.

Such a reductionist approach, however, has immediate and controversial consequences. In the case of time it follows that there can be no time without change. If we seek to treat a period of time as a collection of events, we could not have a temporal vacuum. It would be a contradiction to speak of an interval containing no events if an interval is just a certain collection of events. And in the case of space there could be no motion that was not the motion of bodies in relation to other bodies. Suppose, for example, that we try to entertain the idea that the entire universe is moving to the left. Before and after such a putative motion all bodies would have the same spatial relation to one another. But if locations are to be defined in terms of relations between bodies we cannot have a difference in location without a difference in the relations between bodies. If all bodies keep the same relations to one another it would be a contradiction to speak of them as nonetheless moving. We will return to consider the significance of these consequences after considering the Newtonian alternative to this Leibnizian reductionism.

This alternative might well be called the Cambridge theory — not because Cambridge is the home of lost causes and this is a lost cause, but because the most important articulation of the position was given by the Cambridge neo-Platonists who taught it to Newton. On this view space and time are taken to have an existence independently of the contents of space and time. To put it in theological terms, if it had not pleased God to create the physical world, space and time would nonetheless have existed as the containers into which God could have put events and bodies. Having an existence independent of the existence of events and bodies, the structure of space and time would not depend on the properties of such items. This absolutist or substantialist conception has resulted in the thought that the properties of space and time are to be discovered not by investigations of the physical world but by *a priori* philosophical reflection.

The strains of such a conception are to be found in Newton's writings. This Newton, who I will call *Bad Newton*, held that space and time are aspects of God. They had to be aspects of God because otherwise we would have other infinite things besides God and that would detract from his glory. And since God existed necessarily, these aspects of him had to similarly have necessary existence. Necessarily existing they would have what properties they had prior to his creation of the physical world. Consequently, we do not need telescopes to investigate space and time but the ample armchairs of philosophy and theology.

There can be discerned in Newton's writings a significantly more plausible version of this substantialist view which is not at all a lost cause. To make it clear that this is more suggested than explicitly stated by Newton I will attribute it to a fictional character to be called *Good Newton*. *Good Newton* argues that reductionism has failed to give satisfactory translations of all assertions about space, spatial items, time and temporal items into assertions referring only to bodies and events. *Good Newton* argues that space and time are systems of spatial and temporal items, respectively, which cannot be treated reductionistically. In order to deal with the events and objects which we experience we posit the existence of space and time each with a certain structure. Just which structures to posit is to be determined by considering which gives us the best description and explanation of the contents of space and time. We do not directly observe the properties of space and time but infer them in the course of developing physical theories. *Good Newton* sees the reductionist as operating with too restricted a version of empiricism. He will draw attention to the fact that in physics we frequently infer the existence of unobservables. For instance, without believing in the possibility of observing free quarks we may nonetheless posit their existence on the grounds that this is required if we are to explain the phenomena of hadronic jets.

This form of substantialism, the substantialism of *Good Newton*, might be called *empirical substantialism* in contrast to the *theological substantialism* of *Bad Newton*. To see how empirical considerations could be deployed in arguing for such a position, it will be instructive to elaborate the thought experiment used by Newton in the *Scholium* to the *Definitions* in his *Principia*. (See Thayer in Further Reading.) To this end imagine that we live in a universe consisting *solely* of two spheres each rather like the earth but joined by a cord. We who live on one of the spheres have discovered Newtonian mechanics and have found that it works well when applied to objects on the surface of our sphere. One day we discover that there is a tension in the cord connecting the spheres and ask for an explanation. Given Newtonian mechanics we know that if two bodies joined by a cord rotate, there will be a tension in the cord due to the effects of centrifugal force. But in the case of the spheres there is no motion with respect to any physical body. For everything except small objects on the surface of our sphere keeps a fixed spatial relation to everything else. Thus, if the spheres are rotating with respect to something, it can only be with respect to absolute space. Since rotation would explain the tension we assume rotation and thereby posit the existence of a substantial space.

The above argument draws attention to circumstances in which it would be entirely respectable from the scientific point of view to posit the existence of a substantial space, a space which cannot be treated reductionistically. That is not to say that our world is such a world. Newton thought it was and if he was mistaken it was an entirely respectable mistake. It is to be noted that the argument even in the circumstances sketched is not compelling. There will no doubt be some counterpart of Mach in our imaginary world seeking to explain the tension by reference to repulsive forces acting between the spheres thereby avoiding a substantial space. But such a move is *ad hoc*. For if we have found that Newtonian mechanics works on the surface of a sphere and have no experience of repulsive forces between massive bodies it would be more reasonable to posit rotation, an application of a familiar idea, rather than posit a new repulsive force for the sole reason of avoiding substantial space.

Newton held that we do live in a substantial space. As we now know, this assumption leads to embarrassing consequences. For it turned out that there is no possibility of distinguishing between rest and uniform motion with respect to such a substantial space. In classical physics we can at best detect accelerations with respect to substantial space. Newton was thus led to posit something, substantial space, which gives rise to unanswerable questions; namely, is such and such a body at rest or in uniform motion with respect to that space? Some philosophers of a positivistic orientation have held that to the extent to which a theory gives rise to utterly unanswerable questions it is meaningless. However, we do not need to make such a controversial assumption in order to feel uncomfortable with this substantialist conception of space in the context of classical physics. For it is certainly methodologically undesirable to adopt theories giving rise to unanswerable questions.

We have seen how empirical consideration can be deployed in certain contexts which would favour the empirical substantialism of *Good Newton*. Philosophical considerations concerning meaning have also played a role in this debate which will be illustrated with regard to time. It was noted that the reductionist was committed to the view that there can be no time without change. For the substantialist of either a theological or an empirical kind empty time is a possibility. And this seems to the reductionist to be just nonsense. In the first place he will draw attention to the fact that we cannot imagine empty time. If by 'imagine' he means 'imagine what it would be like to experience' this is certainly so. We cannot be in the position of noticing that everything has stopped, thinking to ourselves 'Is this ever boring! Absolutely nothing is happening.' For in imagining ourselves thinking this we have to imagine change; namely, change in our mental states. But that something cannot be imagined in this sense does not mean that it will not happen. I cannot, for example, imagine in this sense my own death which nonetheless is highly likely to occur one day.

In addition, a reductionist of a positivistic orientation will claim that one could not have evidence for the existence of a temporal vacuum as a prelude to denying the meaningfulness of any theory involving such a notion. If this means that we could not have direct evidence in the sense of actually experiencing it we can agree but reply that much of physics depends on reference to items of which we cannot have direct experience — photons, fields, quarks and so on. We can reject temporal vacua on these grounds only at the cost of rejecting much of theoretical physics. It is standard practice to argue for hypotheses that cannot be directly tested on the grounds of their explanatory power. To show that the idea of empty time is at least no worse off than much of physics, we can sketch an account of circumstances in which an hypothesis employing that idea would be reasonable on the normal criteria used in adopting scientific hypotheses. In this I follow Newton's strategy by offering a thought experiment. An adequate argument to this effect is the subject for an entire chapter and in the present context I can only indicate how it might begin. For reasons of space I borrow a story from Sydney Shoemaker (see Further Reading).

Suppose we lived in a universe having three regions — A, B and C. We, the As, notice by our clocks that all change ceases in the B region for a year every three years. The Bs and everything else in their region is 'frozen' throughout every third year during which time we cannot enter the region. We notice that the same thing happens to the Cs every four years. Thus far we do not have time without change. We have only made reference to periods of time during which there is no change in certain regions of the universe. During a time when no region is frozen representatives of all regions meet to discuss the situation. Initially the Bs and the Cs decline to accept the idea that they are periodically frozen. For being totally 'frozen' they are not conscious of these periods. However, they come to accept this on the grounds that they notice the other regions to be subject to local 'freezes' and that it would explain something otherwise puzzling. This is that every so often it appears to them as if the other regions of the universe undergo a discontinuous change, a change of the sort that would normally take a year to come about. The Bs and Cs draw our attention to the fact that we are the victims of a local freeze every five years.

Once the inhabitants come to agree that they are subject to periodic local freezes they agree to advance their calendars by a year when prompted by the inhabitants of the other regions. At the meeting someone draws attention to the fact that every sixty years the local 'freezes' will match up giving a year in which there is absolutely no change anywhere in the universe. If this begins at midnight on New Year's Eve we will be able to lift our glasses and drink a toast to the year that went past unnoticed between the time we picked the glasses up and set them down.

As in the case of the Newtonian thought experiment this argument is more persuasive than compelling. For it would be possible to insist that we should adopt more complicated hypotheses concerning the local freezes. For example, we could say that the A region freezes every five years except one year in sixty in which it skips a freeze. We could similarly complicate the hypotheses governing the B and C regions. But why should we insist on the more complicated hypotheses unless we are begging the question at hand by assuming that there simply could not be empty time? Obviously much more would need to be said to make this argument for the respectability in certain contexts of empty time itself respectable. Enough has been said, however, to show that from the philosophical point of view the reductionist has a case to answer.

Against the background of classical physics it seems that the container view of space and time of *Good Newton* has more to commend it than the contents view of Leibniz in certain contexts. For we have seen that the substantialist's space would explain certain observations better than a reductionist's could and we have seen that the reductionist's assumption that empty time is nonsense is dubious. It is often claimed that this conclusion holds at best in the context of classical physics. The positivists, for example, claimed that Einstein's *Special Theory of Relativity* was a victory for the Leibnizian approach over the Newtonian approach. However, attention to some of the basic features of the Special Theory of Relativity shows that this conclusion is not warranted.

Newton's container view encourages the idea that any event E_1 is either simultaneous with any other event E_2 or it is not. For the location in the time container of E_1 is either the same location as the location of E_2 or it is not. The discovery that there is an upper maximum to the speed with which information can be transmitted, the speed of light, means that we cannot find out which event at our present location really is simultaneous with some distant event. In fig. 3.1 a light signal is represented as leaving our location at time t_1 to arrive at the occurrence of the distant event E_2 and is reflected to return to our location at time t_2. The event E_1 occurring at our location which is simultaneous with E_2 could be anywhere between t_1 and t_2 If we insist that there is some event which is *really* simultaneous with E_2 we have the embarrassment of allowing questions which do not admit of answers. Einstein proposed dropping the notion of absolute simultaneity to avoid this consequence. Instead we should talk of simultaneity relative to a particular frame of reference. Assuming we adopt the convention of fixing the time of E_1 as midway between t_1 and t_2 we find according to the Special Theory that in other frames of reference in motion relative to ours E_1 and E_2 are not counted as simultaneous. In general different frames of

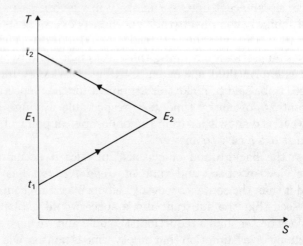

Fig. 3.1

reference in relative motion will assign different values to the time and the spatial distance between the same pair of events. While these quantities vary from frame to frame there is in the Special Theory a quantity that does not vary. This is space—time distance between a pair of events defined by the following expression: $(x-x')^2 + (y-y')^2 + (z-z')^2 - c^2(t-t')^2$, where c is the speed of light and the events occur at times t and t' and at spatial locations (x,y,z) and (x',y',z'). In a different frame of reference the spatial and the temporal separation between these events will be different but these will vary so as to give the same space—time separation. The fact that while spatial and temporal separation taken separately vary, and space—time separation does not, suggested to Minkowski that the fundamental item of concern should not be space and time treated separately but space—time:

The views of space and time which I wish to lay before you have sprung from the soil of experimental physics, and therein lies their strength. They are radical. Henceforth space by itself, and time by itself, are doomed to fade away into mere shadows, and only a kind of union of the two will preserve an independent reality (see Further Reading).

In explicating the notion of space—time it will be convenient to use 'container' terminology. Space—time is a system of space—time points in which events are located. A point of space—time is a point of space at a moment of time. The full-stop ending the preceding sentence is at a different location in space—time at each different moment of time. The full-stop's history would be represented as a line of space—time, a space—time having four dimensions, three spatial and one temporal. It is, however, important to note that the spatial and temporal dimensions of this space—time are treated as being very different, a difference which is signalled by the fact that there is a different sign before the temporal component in the expression for space—time separation.

When Minkowski put space and time together this was not like putting a knife and fork together on a table, it was more akin to putting an acid and base together. Space and time put together in the context of the Special Theory of Relativity gives us something with new properties, something that does not simply come apart into space and time as traditionally conceived. In classical physics spatial and temporal separation taken on their own were invariant in the sense of being the same in all frames of reference. As was noted above, in the Special Theory this is no longer the case. The invariant is instead space—time separation. And what lies behind Minkowski's claim that space and time must fade away in favour of space—time is a philosophical principle according to which what is really real is what is invariant.

What would Newton and Leibniz make of this move to space—time? Let us imagine that the current Princess of Wales having read Minkowski finds herself travelling back to 1715 thanks to Dr Who's Tardis. She inquires of Newton whether Mr Leibniz was not after all nearer the truth. Newton, I imagine, would reply:

'Not at all. Einstein and Minkowski have vindicated what I always thought. Of course I was wrong about the details. I thought I could treat space and time *separately* and that is why I advanced substantialist characterizations of each. Now I see that it is physically more reasonable to admit but one container, one substantial item, space—time. This Minkowskian space—time exists independently of its contents.'

Crossing the Channel the Princess puts to Leibniz the thought that the use of space—time in twentieth-century physics is more in keeping with the substantialist position of Newton. Not at all, Leibniz could reply:

'Things have worked out as I thought they would. True I got the details wrong. I assumed that I could *separately* reduce all talk about space to talk about bodies and all talk about time to talk about

events. The incomparable Mr Einstein shows that we must use
space — time rather than space and time. So I will have to reduce all
talk about that to talk about, say, point   events. Of course I must
confess to a temporary problem: empty space — times. But that is
nothing new and does not arise because of any special features of
twentieth-century physics. My critics used to worry me about it in
the context of classical physics. I very much doubt that such nonsense
will be a permanent feature of gravitational theories. What is more,
certain contemporary philosophers of language are advancing
verificationist theories of meaning which should help to establish the
nonsensical character of talk about empty space — time.'

My suggestion is then that the Special Theory of Relativity is
neutral in the debate between the substantialist and the reductionist.
That being so the question arises as to whether the General Theory
of Relativity provides a resolution. This is a complex and
controversial matter which can only be touched on within the confines
of this chapter. For some this brings definitive victory to Newton.
Solutions of the field equations of the theory give descriptions of the
structure of a space — time. And these can be solved even when it is
assumed that there is no matter or radiation whatsoever. But if
space — time could not exist in the absence of matter as Leibniz
would have us suppose we should expect something to go wrong
when we attempt to find out what an empty space — time would be
like. Such is not the case. The theory admits of the possibility of the
existence of different well-behaved, entirely empty space — times.
Those of a reductionist orientation have only been momentarily
silenced by this result. Drawing attention to the fact that, in general,
mathematical equations may have some solutions which are
physically sensible and some that are not, the reductionist argues
that they are extraneous roots which should be cast out. Rather than
establishing the possibility of empty space — time they show a
deficiency in the General Theory. And other reductionists have
sought to modify the field equations so as to block vacuum solutions.

Whether the General Theory of Relativity favours reductionism or
substantialism or is neutral between them is an open question. The
only conclusion that can be drawn with confidence is that there will
be no purely philosophical nor a purely physical resolution of this
controversy. In the exchanges between Leibniz and Newton against
the background of classical physics such considerations intermingle.
Reference is made to the more scientific question of which theory will
provide the better explanation and reference is made to the more

philosophical question of the meaningfulness of, for example, rotation relative to absolute space. Then as now, these conflicting orientations to space—time must be tested by reference to both physical theories about gravitation and philosophical theories about meaning. Thus far it remains an open issue.

In seeking to relieve our perplexity about what we mean by 'space' and 'time' we considered theories about the relation between the physical world and space and time. These theories offered conflicting elucidations of *an* aspect of our conceptions of space and time. In seeking to adjudicate between them we have seen that reference must be made to physical theories. This suggests that no satisfactory philosophical understanding of the nature of space, time or space— time can be achieved by remaining at the purely semantical level. Our question about meaning took us to physics and certain results in physics take us back to meaning. We have not said in plain words what we mean by 'space', 'time' or 'space—time' and remain, in Barrow's eyes, but 'Quacks'. Still, to have an appreciation of an issue that needs resolving may be progress of a sort.

FURTHER READING

Barrow, I., *Lectiones Geometricae*, quoted from translation given in *Concepts of Space and Time: their Structure and their Development*, ed. M. Capek (Reidel, Dordrecht, 1975), pp. 203—4.

Leibniz, G. W., *The Leibniz—Clarke Correspondence*, tr. H. G. Alexander (Manchester University Press, Manchester, 1956).

Minkowski, H., 'Space and Time', in *The Principle of Relativity*, ed. W. Peerrett and G. B. Jeffrey (Dover, New York, 1923).

Nerlich, G., *The Shape of Space* (Cambridge University Press, Cambridge, 1976).

Newton-Smith, W. H., *The Structure of Time* (Routledge and Kegan Paul, London, 1980).

Shoemaker, S., 'Time without Change', *Journal of Philosophy*, 66 (1969), pp. 363—81. There are difficulties in Shoemaker's story which I have tried to rectify in my book, *The Structure of Time* (see above).

Sklar, L., *Space, Time and Spacetime* (University of California Press, Berkeley, 1974).

Swinburne, R., *Space and Time* (Macmillan, London, 1968).

Thayer, H. S., (ed.), *Newton's Philosophy of Nature* (Hafner, London, 1953).

Van Frassen, B. C., *An Introduction to the Philosophy of Time and Space* (Random House, New York, 1970).

4

Big Bangs, Black Holes and 'Time's Arrow'

ROGER PENROSE

What I have to say relates to the beginning and end of the universe; but also to perfectly commonplace events, such as what happens when a cup of coffee is stirred. With regard to the beginning of the universe, some of my remarks may be a little unfashionable. That is to say, I shall not talk about the current ideas of 'inflationary universes' (with which I am somewhat out of sympathy). What I shall be concerned with is the seeming paradox that the laws of physics, or such of them as we understand, appear to be reversible in time, whereas we are constantly confronted with large-scale phenomena that are manifestly not reversible. Think, for example, of atoms colliding with each other. The laws which govern the behaviour of such small particles are completely reversible in time, yet the macroscopic objects which are composed of vast numbers of these tiny objects behave in ways which are patently irreversible. And that applies to such things as people and so on. Take a film of almost any human activity, run it backwards, and what you see corresponds to nothing that one would, or normally could, encounter in real life.

There is, of course, a kind of pat answer one can make to this, and that is to appeal to the second law of thermodynamics. This law says that there is a thing called *entropy* and that this entropy goes up as time goes on. We postulate that. Now that *is* an irreversible law. But it's a law which applies to large systems of particles, not to individual particles. It says, roughly speaking, that if things are peculiar early on, in certain manifest ways, then they will get less and less peculiar as time passes. A typical example is a box containing gas, which starts off with all the gas up in one corner. If you leave it, the gas will spread itself uniformly all over the box. When the gas is all in one corner, the system is in a low entropy state; and when, by contrast, the gas is uniformly distributed with a certain very specific distribution of velocities and so on, it is in a high entropy state. Thus it evolves from the low entropy to the high entropy state.

Actually there is nothing very remarkable about the fact that the entropy of a system goes up, because all that really means is that if you start a system off in a very special state and you let it go, it is likely to end up in a less special state. It may be useful here to introduce the notion of what is called *phase space*. Phase space is an imaginary space with a large number of dimensions, which is used to describe a physical system. The state of the system, at any given time, will correspond to a single point in that space. (In classical mechanics, a system of N point particles can be represented by a space of $6N$ dimensions, six for each particle, corresponding to its position and its momentum with respect to the x, y and z axes, respectively.) If we consider a phase space for our box containing gas, then there will be certain regions of this space which are manifestly peculiar — the gas being all in one corner, for instance. Such regions of phase space will be very small. But there are other states which look very much the same as each other, such as the states in which the gas is pretty uniformly spread out all over the interior of the box. These states will jointly occupy a much larger region of the phase space. Clearly, if you start the state off in a manifestly peculiar region of phase space, and you just let the state evolve, so that the phase-space point wanders around somehow, the chances are that it's going to get into a large region.

That is more or less what the second law of thermodynamics says: namely, that if you start in some tiny region of phase space that has been specially singled out, or something like that, then you will soon find yourself in a much larger region. This thing called entropy, which is usually denoted by the letter S, is, roughly speaking, the logarithm of the size of the region of phase space in question — a region singled out in terms of its manifest macroscopic properties. One thing that should be made clear is that there is an absolutely ridiculous difference in the sizes of these different phase-space volumes. (This is one reason why one takes the logarithm. It's not the main reason, but one advantage of taking logarithms is that the differences then don't look so extraordinarily big.) If, on the other hand, you are just told the entropy values corresponding to two regions of phase space, then you have to take the opposite of logarithms, namely exponentials (that is, raise to the power of those numbers) to find out the relative sizes of the phase-space regions.

There's nothing at all remarkable about the fact that the entropy goes up in the future. Nevertheless, there *is* a very peculiar thing about the second law of thermodynamics. It actually is a most extraordinary physical fact, and what's extraordinary about it is that if you look into the *past* instead, you find that things don't behave in this apparently 'probable' way. Let's take Newtonian physics; forget about quantum mechanics for the moment. If you take standard Newtonian physics, the laws are time-reversible, and you can just as well evolve into the past as you can into the future. Both are equally allowable. So let's go into the past. If we evolve the state into the past, we find that, because the second law says the entropy's going up in the normal direction, the entropy is going down as we go into the past. You start with a point in some region in phase space, and you find that as you evolve into the past, the point goes into more and more peculiar regions as time proceeds (backwards).

This may not seem particularly remarkable, but only because we are used to thinking of things evolving forwards in time. Didn't I say that the second law of thermodynamics is a time-irreversible law? Well yes. It is, if it's stated in the simple form: entropy always increases in the direction earlier to later. But when this is explained along the lines sketched above, as simply a statistical consequence of the ordinary reversible laws of physics, as applied to very large aggregates of particles, the second law seems to be being deduced from something time-reversible! Suppose you take a closed system (free, that is, from external interference), examine it at time T, and find that its entropy is less than the maximum possible for the system. Then statistical mechanics tells you that, if you examine it at any time later than T, the chances are that it will be in a state of higher entropy. But, in the absence of other information, save what the state was like at time T, the very same argument says that if you examine it at any earlier time, it is likewise probable that it will be found to be in a state of higher entropy! Why, then, don't we actually find (or expect to find) such behaviour in the majority of cases? The answer is that we interpret the state of relatively low entropy as being a consequence of it having been started up in a state of even lower entropy, not of a random fluctuation away from a state of high entropy. As the jargon has it, the temporal asymmetry derives not from the evolution laws so much as from the assumed *boundary conditions* in the past: the assumptions we make about how things started off.

So the puzzle is: why was the entropy so low in the old days? Well, when were the old days? Clearly, entropy is increasing now, and has been for a very long time. The question, then, is how far back one is entitled to extrapolate that trend. Well, physicists usually just assume the second law. They don't think that at a certain time the rules changed, and before that entropy was going the other way. So we must take it that it just keeps on going down as we go further and further back in time. It looks, therefore, as though we might as well go right on back until we get to the beginning: time zero. Well, there is supposed to *be* a beginning on the modern view of cosmology — that is, the big bang. And in fact there is quite a lot of evidence from various quarters that there was a highly condensed state in the early stages of the universe. There is evidence both from theory and from observation that there was some such thing as the big bang, that the universe was originally concentrated in some very tiny region. The entropy must presumably have been at its lowest value at the big bang.

That's fine. But, on the other hand, there is a sort of paradox involved here. I said there was quite a lot of evidence in favour of the big bang. Some of that evidence comes from detailed calculations that astrophysicists carry out, when they think about the early stages of the universe. And one of the prime ingredients which goes into these detailed calculations is that the universe, in its very beginnings, was in a state of *thermal equilibrium*. Now, thermal equilibrium — what does that mean? That means that you're in a state of maximum entropy. Well you don't have to be a mathematician or physicist to realize that there's something a little bit puzzling here. The entropy has gone down and down and down in the past, until it reached its maximum value!

I've been puzzled by this for quite a long time — very many years. I've often mentioned it to my astrophysicist friends and I've got this kind of strange look usually. I felt that I must have said something stupid. People obviously thought that I'd somehow missed something — some essential ingredient in this whole argument. Well I suppose the essential ingredient that I'm supposed to have missed here was the fact that the universe is expanding. The idea seemed to be that one starts from this very small state, the universe gets bigger and for some reason the entropy that's allowed in the early stages is very, very tiny. The universe was using up all the available randomness, if you like, and being as random as it could, but since the universe was so small, the possible maximum entropy was much lower than it is in the universe today. Well, that perhaps kept me quiet for a little bit; I'm not sure. But, after thinking about that for a while, I realized that it doesn't make any sense. It doesn't make any sense for a perfectly obvious reason. And one thing that worries me about this whole discussion, actually, is the fact that it only involves ingredients which are perfectly obvious. How is it that one can come to a conclusion with which everybody seems to disagree, when you're only putting in things which are perfectly obvious?

So why doesn't this make sense? Well, it might be the case — people don't know for sure — that the universe is closed, finite, a kind of three-dimensional analogue of the surface of a sphere. That's one possibility. I should say that the evidence, for what it's worth, is probably slightly in the other direction. But that's not the point, because clearly it *might* be the case that the universe is closed. So let's suppose that the universe *is* closed. Then what will happen? It will reach its maximum expansion, turn around, come back again and will then finally end up in what's sometimes referred to as the *big crunch*.

Now the suggestion we're currently exploring — the standard response to my puzzle — is that the universe indeed started in a state of thermal equilibrium; but that that still corresponded to a very low total entropy, because the small size of the universe placed a very low ceiling on the maximum entropy it could have. For a system to be in a state of thermal equilibrium just means its having the highest entropy possible, given its size. So what happens, on this view, as the universe expands? Well very early on — so the story goes — the universe is expanding very rapidly, so rapidly that the expansion rate exceeds the rate at which the system is able to evolve back to a state of equilibrium, in obedience to the second law. So at first the universe is getting further and further away from thermal equilibrium. As the rate of expansion slows, however, the rate at which the universe approaches thermal equilibrium once more exceeds the expansion rate. Well, that may initially strike one as quite reasonable. But if this general line of thought is correct, what is supposed to happen to the entropy when, as we're now assuming, the universe begins to contract? On the one hand, the second law of thermodynamics says the entropy must keep right on going up. But, on the other hand, the view I've just sketched clearly entails that it *can't* go on doing that indefinitely. For as the universe continues to contract, the maximum possible entropy will get progressively smaller: and you can't get a quart of entropy into a pint pot, so to speak. We appear to have a contradiction.

Well, there is a view on this subject which I should perhaps mention. But I don't want to spend too much time on it, because I don't believe it myself and I think one can present a strong argument against it. The suggestion is that perhaps somewhere in the middle of its history, the universe actually reaches its maximum entropy, which then starts going down again, as the universe contracts. This may sound plausible. But you have to worry about what it actually entails. Current thinking has it that it is entropy that determines the arrow of time. It is highly plausible that the macroscopic asymmetry of time is grounded in the second law of thermodynamics; that it is that which ultimately explains such things as why we can remember the past but not the future. If that's correct, then what makes us think of one direction in time as 'earlier' and the other as 'later' is that the one direction corresponds to decreasing and the other to increasing entropy. So if entropy were to stop increasing and start decreasing again, it would seem to follow that time itself would come to have a different directional sense. Things after the turnaround would behave as though the future lay in the direction we think of as the past. And that gives rise to all sorts of problems. One problem is that the universe starts to recollapse, for instance, before you see it recollapse and the stars are all coming towards you. So what does the light do when it starts going the wrong way? What exactly is supposed to happen to light rays that span the transition between the period at which entropy is increasing and that at which it is decreasing? Will light rays diverge from a source prior to the turnaround, and then miraculously reconverge on some point after the turnaround — a point which, from the post-turnaround perspective, constitutes their source?

We also have trouble with various paradoxes. Suppose, for example, you have people living after the turnaround, in the contraction era. We can call them the *contraction people* (though they, of course, will think of us as the contraction people, and themselves as expansion people). They, presumably, will remember things that happened in what, from our perspective, is their future. If they were to be told things that had happened in what we would regard as their past, this would create a paradox. Because what would then prevent them from choosing to do something different? But now *we* are in what they regard as their future. So why couldn't we send the contraction people a message, telling them about something they would regard as in the future: something, moreover, which it would seem to be in their power to do something about? What we could do, in principle, is put a lot of information in a sealed container and simply let it loose in the universe. If current estimates about the size of the universe are correct, there wouldn't be much difficulty, in principle, in building some box that would last through to the contraction phase, on the assumption that the universe will eventually start to contract. There is then no obvious reason why the contraction people shouldn't come across this container, find out what was going to happen in their future and do something to prevent it happening. Suppose, for example, they were to send out a space probe of their own on such a course that it would rendezvous with the container our side of the turnaround and destroy it, thereby preventing them from ever receiving the container in the first place? Actually, things are not quite as simple as I've made them sound. But there are, in principle, very severe difficulties with this kind of picture.

But you might say to me: 'All right, then, perhaps the universe isn't closed.' And, after all, as I said, the evidence, for what it's worth, is perhaps somewhat on the other side, suggesting that the universe may well continue to expand indefinitely — in which case these problems won't arise. Well, that doesn't really get you out of the difficulty. And this is where the black hole part of the title comes in.

At this point it will be useful to draw a diagram. It's quite helpful, sometimes, to draw pictures of the whole universe on a blackboard or on the back of an envelope or something like that. Here we can make use of the sort of devices employed in certain map projections, where we don't mind distorting distances, but don't wish to distort angles. What I am talking about here are *space—time* diagrams; and the idea is to avoid distorting the *light cones* too much. A light cone is an imaginary surface, associated with a point in space—time, comprising the paths of all possible light rays that pass through that point. Time progresses from the lower to the upper part of the diagram, and a particle, or any other object, is represented by a curve referred to as its *world-line*: this gives the history of that particle as time progresses. The light cone is the important ingredient here, since it shows you how signals can propagate. Particles necessarily have their paths constrained to lie within the light cones: if you take any point in space—time, and take the light cone associated with it, then any world-line that passes through that point must lie within the corresponding light cone. This is just another way of saying that no particle can travel faster than the speed of light, or more precisely, that no particle can exceed the *local* light velocity. Now in general relativity, you may have light cones drawn all higgledy-piggledy and so on (provided there is a smooth transition from the light cone you get at any one point to that which you get at any other). But the essential thing is that if you consider the local observers travelling around on their world-lines (if that's the right way of talking — strictly speaking they *are* their world-lines), then these world-lines have to be constrained, wherever they are, to be within the light cones. So if you think of it locally, time will be going in the direction from the backward to the forward part of the light cone. No-one is allowed to cross the boundary of his own light cone. So that's the sort of picture one has. And if you imagine these cones being drawn on a rubber sheet, then general relativity allows you to stretch and bend them around locally, if you like. What you get will still obey the same fundamental laws.

What you can do, though, is draw pictures where these light cones, rather than looking distorted or leaning over at crazy angles and so on, all sit up in a fairly orderly fashion. Let's start with the more standard type of space—time diagram. We can, for example, represent the standard cosmological models in quite a useful way. The closed universe that I was talking about just now starts off with a big bang — at least, that's the way people usually think of it. The big bang seems to be a singularity. That's where all the curvatures go to infinity and so on. So it can be represented as a point at the bottom, while the big crunch is represented as a corresponding point at the top. It's useful, here, to throw away two dimensions and think of the universe as having just one space and one time dimension for the purposes of visualization. So we can think of it as the surface of a sausage of some sort, with space going around the circumference and time running lengthwise from the bottom to the top (see fig. 4.1).

That's one way of representing the history of a closed universe. Here the light cones are all arranged lengthwise along the surface. But then they have a habit of getting rather squashed when you approach either end. And to see what's happening in detail in the vicinity of the big bang and the big crunch is consequently rather difficult with this picture. What's useful, therefore, is to stretch this

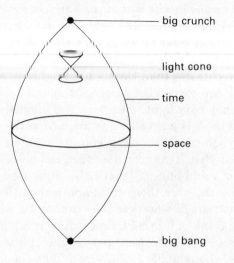

Fig. 4.1

model at the bottom and the top, and open it out like a map. When it's unfolded it looks something like fig. 4.2. This represents the whole universe. The lower and upper edges correspond to the big bang and the big crunch, respectively. And the light cones are now drawn all the time at 45°, no longer becoming distorted as we approach the two singularities. That's the advantage of this form of diagram; and it makes for much clearer representation of the important features. All the standard closed universes postulated by cosmologists may be depicted in this kind of way.

Imagine now a particle emerging from the big bang, and somebody sitting on it looking out at the universe. One thing you will find is that, as time goes on, more and more of the universe is encompassed in his field of vision. At any one time, only a certain set of particles will be visible to the observer. The boundary of this set is called his *particle horizon*. At any given time he can, in principle, only see a certain region of the universe. A particle that lies on the far side of his particle horizon is one that has not yet come into view. But if he waits a bit longer (always assuming he lives long enough) he will eventually see it. This is the feature of a particle horizon. Now you'll notice that the big bang is, in fig. 4.2, stretched out into a whole space-like region. This brings me to something that often worries people: something which starts off at point *A*, say, can have, initially, no influence on something that starts off at point *B*, because *A* lies

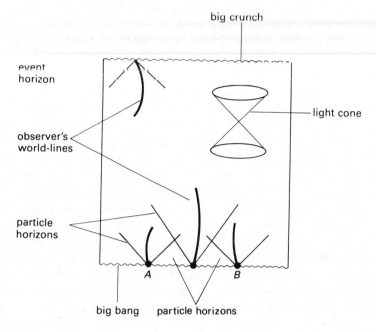

Fig. 4.2

outside the light cone, and hence outside the particle horizon, of *B*. A lot of difficulties that people have with the standard cosmological pictures result from these so-called *horizon problems* in cosmology. There is, for example, a lot of activity these days devoted to what are called *inflationary models* of the universe. And one of the main aims of these models is to try to remove this seeming difficulty that one has in thinking about the early stages of the universe, and to try to overcome these horizon problems. But I don't want to say anything more about these inflationary models here, other than pointing out that, even though these ideas of mine were developed before inflationary models became fashionable or even known about, the inflationary models don't in any significant way affect the present argument.

There is also another kind of horizon that I should mention (also shown in fig. 4.2). If you have an observer approaching the big crunch, there are some events which are forever invisible to him: these are said to lie beyond his *event horizon*. By the time this observer is finally crushed out of existence, these events will still not have been observed. Had the observer chosen to do something else at a sufficiently early stage, so that his world-line went in a different direction, then he would have been able to observe some of these events. But then there would have been other events which he would have been completely unable to observe: events which, as things are, he can observe.

Event horizons thus stand to the big crunch as particle horizons stand to the big bang. The possession of such horizons is a standard feature of these models. I mention them only because they keep coming up and because I wanted to draw you a picture of a black hole in this kind of scheme. There will originally have been a star, galaxy or something, which collapsed under its own weight, to the point where its own gravity became sufficiently intense to overcome all opposing forces. This collapsing matter is shown at the bottom of fig. 4.3. In the middle there is a somewhat unpleasant singularity. Now the point is that the light cones are, so to speak, pulled in by the gravitational field of the collapsed matter. The event horizon represents the point beyond which it is impossible for light to escape: points on the horizon thus have their light cones tangential to the horizon. Once inside the region bounded by the event horizon, an observer cannot escape, because to get back across the horizon it would be necessary for his world-line to pass outside his own light cone: for it to go in what is called a *space-like* direction. And this is not allowed. So all he can do is go from the outside to the inside. He cannot turn round and go back again without exceeding the local light velocity.

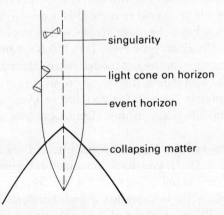

Fig. 4.3

So that's what a black hole looks like, drawn in the standard way. But that is not always the most useful way to draw it. In fig. 4.3 the singularity is shown as a vertical line, and the light cones are all falling in towards it. If, in accordance with our new type of diagram, I turn the light cones round so they look upright, then you find that the singularity becomes horizontal (see fig. 4.4). The event horizon is now shown as a line at 45° to the vertical. Then out to the right-hand side you've got infinity, which can be included if you like. I shouldn't worry too much about that. And on the left we have the axis of symmetry. You should think of the whole thing rotated round that axis. In reality, it's centred on the dotted line.

Now suppose we have a brave astronaut who decides to fly his spacecraft into a black hole in order to carry out various experiments in connection with the second law of thermodynamics. He keeps taking glasses of water and smashes them on the floor. Several rather crude experiments, I'm afraid. But then he happens to be a rather crude astronaut!

I'm still addressing this original worry of mine. Suppose it were true that the reason that the entropy is going up is merely that the universe is expanding. Well here in the black hole, you have locally a universe which is collapsing about you. The future of our exceedingly brave astronaut is dominated by the singularity. Now suppose you're taking the view that whatever happens to the universe is governed

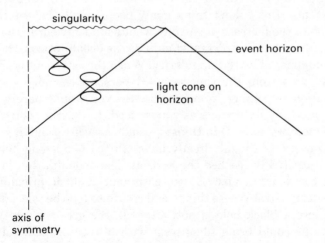

Fig. 4.4

by its boundary conditions. Suppose, further, that you're not going to be biased as regards the directions forward into the future or back into the past; and seeing that the laws of physics usually are time-symmetric, there's no reason to be biased that way. That means, then, that whatever happens to that astronaut's entropy should be governed by the singularity. His universe is collapsing completely. According to the hypothetical view we're now considering, entropy would go down as the singularity is approached. We should therefore look at things the other way around, for the purposes of applying the second law. Things inside the horizon would behave as though, from their own point of view, this region of the universe was expanding. Regarded from this time-reversed perspective, there is no information coming from anywhere but the direction of the singularity: what we think of as an event horizon acts as a particle horizon. Somehow, according to this view, the astronaut seems to find entropy going into reverse as he crosses into the interior of the black hole.

Well, I don't believe it. And I don't think many people would. In principle this is an experiment that could actually be carried out, because there is very possibly a large black hole in the centre of our galaxy. Of course it's not something one would do lightly: it would cost a lot of money for one thing. But in principle, there is no reason why one shouldn't travel to the galactic centre and enter this black hole. It would leave a certain amount of time, perhaps a few hours. At any rate, for *certain* galaxies, it would be a number of days. So you would then certainly have sufficient time to conduct some experiments. But I don't think many people really believe that the direction of the increase of entropy would suddenly turn around.

In fact, even when you're just outside the event horizon, the future is still dominated by the singularity. When the astronaut is only just outside, then from the time-reversed perspective, only very little information is coming from outside. So you would hardly expect things *suddenly* to flip over as you crossed the horizon: you'd expect a little more warning. Odd things — such as self-assembling broken water glasses — should already have started happening before the astronaut actually reached the horizon. They should, that is, if you believe this point of view. So our astronaut ought in principle to be able to demonstrate these things and return to tell the tale. (Actually falling into a black hole is not, after all, the greatest way to die.) Perhaps he could bring films back with him to prove it (although people might be sceptical as to which way the film had been running). But, as I say, I don't really think such an entropy reversal would take place. I think the entropy would go right on going up, and would continue to do so as the astronaut crossed the event horizon. There's no reason why it shouldn't and every reason to believe that it would.

So what's wrong with the original argument which suggests that there is a low entropy ceiling for a small universe? What's wrong, I believe, is the assumption that our state of seeming thermal equilibrium actually represents a state of maximum entropy. Not only was the universe *not*, in my view, in a state of maximum entropy at the big bang; I think it was in a state of quite absurdly low entropy relative to what it 'might' have been.

The question to consider is what actually happens to the universe, in detail, as it, or part of it, eventually stops expanding and begins to collapse. What sort of end state do you get? Well, you get something which isn't remotely like the state at the start. In fact there already exists a reasonable-looking picture of what the universe would be like in the last stages of the collapse. First the galactic centres collapse, sucking in surrounding stars and gas. And as they start coming together, these galaxies congeal with each other. Their black holes fuse to form a larger and larger congealed mass. The singularity at the end is a great mess. It's nothing like the singularity at the beginning, which seems to have been extraordinarily uniform and symmetrical.

The initial singularity is sometimes said to be a place where one hasn't the remotest idea what's going on in the physics. Well in one sense that's fair enough. But on the other hand, there's a tremendous amount of structure in the big-bang singularity. It's not at all, in its geometric form, a generic thing. By contrast, one expects the final state to be some very general type of singularity. And there's a great deal of difference. In fact, one can make some kind of estimate as to the difference of entropy owing to these congealing black holes in the end, and compare it with more everyday entropy values and with the entropy at the beginning. I'll just give you a few figures. Let's think about a closed universe. And let's suppose it to contain 10^{80} baryons: protons and neutrons that is. Why do I take that figure? Well, most people take this figure. The reason, I think, is that Eddington at one time produced a marvellous number which was supposed to be the exact number of protons in the universe. And although nobody any more believes Eddington's theory, they still use his figure for the number of baryons in the universe. That's the only reason I can think of, because there isn't really much evidence to support it. We do know that it is at least that sort of size. But it might be much huger. If it's bigger, then things are even more extreme than I am about to suggest. Do if you like, I'm taking a very conservative view.

The first line in table 4.1 gives the entropy associated with the big bang. I have put that down because, in the discussion about the early universe, there is a figure which is sometimes bandied about, with good reason, which is either 10^8 or 10^9 — something of that order. This is supposed to be the observed entropy per baryon (using units for which Boltzmann's constant is taken as unity). Now the observed value is not necessarily the actual entropy value of the big bang. It is the value that one obtains from the *background radiation*. There is this radiation out in space, which is, as it were, left over from the hot big bang. A lot of the evidence for the existence of the big bang comes from this background radiation. Now there are a great many photons for every baryon. And since the photons have lots of degrees of freedom, that gives an entropy per baryon value, which is, roughly speaking, the number of photons that there are for every baryon. (Recall that in connection with the discussion of the dimension of phase space it was noted that any particle — and this includes photons as well as baryons — has six degrees of freedom. These can be described as three for its position, at any one time, two for its direction of motion and one for its energy. For N particles there would be $6N$ degrees of freedom. For randomly moving independent particles the entropy would be essentially equal to the number of degrees of freedom. So with 10^8 times as many photons as baryons, the entropy would be dominated by the photon number, and therefore about 10^8 times the baryon number.)

Table 4.1

	S/B	S
'Observed'	10^8	10^{88}
1 $M_\odot$ bh	10^{20}	10^{100}
10^6 $M_\odot$ bh in 10^{11} $M_\odot$ galaxies	10^{21}	10^{101}
10^{11} $M_\odot$ bh	10^{31}	10^{111}
Collapsed universe	10^{43}	10^{123}

S, total entropy; S/B, entropy per baryon; $M_\odot$, one solar mass; bh, black hole(s)

People have been trying to explain why this 'observed' figure should be so huge — because it's much larger than the sorts of entropies one finds in ordinary physical processes. I've also put down the total entropy, which is quite useful. The total entropy is 10^{80}. But is that figure really so big? That's what I want to ask. Let's think a little bit about what is the right way to look at this. I talked earlier about phase space. Figure 4.5 is supposed to be a phase space for the whole universe. (I'm afraid you'll have to use your imagination here.) Every point in this picture represents a possible universe. It doesn't make much difference what stage of the universe's evolution you consider. But if you like, you could think of it as the way the universe started off. Imagine the poor Creator trying to start the universe off in a way that would result in something like what we see. Then it's a question of how carefully the Creator has to aim in order to hit a point which would give us a universe looking something like the one we actually find ourselves in. This is the question I want to address next.

Now I shall put it to you that this is actually a ridiculously small region to aim for. That's the key thing. I said earlier that my argument was going to be based mainly on things that were perfectly obvious. But in fact the thing that I want to say next is by no means

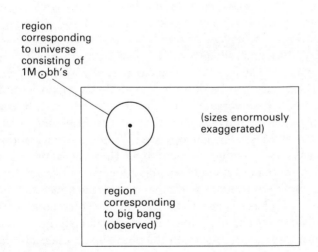

Fig. 4.5

obvious. Bekenstein and Hawking (see Further Reading) have given the following formula for the entropy S_{bh} of a black hole:

$$S_{bh} = \frac{kAc^0}{4\hbar G}$$

This is a very beautiful formula. Most of the symbols in it are actually irrelevant to what I want to say. The only things on the right that are relevant are A and perhaps 4. k is just a constant: Boltzmann's constant; c is the velocity of light, $\hbar$ is Planck's constant divided by 2π and G is Newton's gravitational constant. If you're a relativist like me, you simply put all those things equal to one. You can get away with that (though you have to observe a certain amount of caution: you can't put 4 equal to one!). But still, the essential formula is $A/4$. A is the area of the black hole, the area of the surface of the event horizon. If you know that area, this formula tells you the entropy that should be associated with the geometrical configuration of the black hole. There are various reasons for believing in it, which I don't want to go into; they derive from different points of view. Bekenstein approached the matter by way of a thought experiment, having to do with what would happen if you lowered things into a black hole. Hawking's approach was based on completely different considerations which arise when you do quantum-field theory in curved spaces. And, lo and behold, these two approaches came together. Bekenstein managed to get the A and Hawking added the 4 (Hawking's achievement was by no means the lesser one).

Anyway, it's a marvellous formula. Now let's put some figures into it. When you put the figures in you can actually work out how much entropy there ought to be in a black hole. Let's suppose, first, that every star in the universe was not a star like the Sun, but a black hole, with a mass equivalent to that of the Sun: one solar mass (symbolized as $M_\odot$). The entropy we then get per baryon for these black holes is 10^{20} — huger, much huger than the figure we started with, by an enormous amount. So, going back to our earlier question, a Creator aiming at random would have been much more likely to hit this region of phase space than the region corresponding to the actual universe (at the big bang). It's a ridiculously huger region. I have shown these as two spots on the phase-space diagram (fig. 4.5). But in fact I've had to exaggerate grossly their sizes, particularly that of the smaller one: drawn to scale it would be much smaller than microscopically small, by comparison. It's much easier to hit the larger region (with the black holes) than the smaller one (the 'observed' one). I'll tell you by how much. If the whole universe were built out of these solar mass black holes we'd have a total entropy of 10^{100}. This means that it's $10^{10^{100}}$ times as likely that the larger region would be hit than the smaller one!

But you'll retort, of course, that the universe isn't made up of solar mass black holes. That's true. But on the other hand, there might very well be lots of 10^6 solar mass black holes in 10^{11} solar mass galaxies. These would be very large black holes, by ordinary standards. And you gain in entropy by having large black holes. Many galaxies probably do have large black holes at their centres, even though there probably isn't a very large one at the centre of our own galaxy. Having a fairly sizeable black hole in the centre more than compensates for the rest of the matter being in the form of stars. So this would correspond to a larger region still of phase space. It would have been far easier for the Creator to have hit *this* region. Well, this may still be fairly unrealistic. The actual present entropy could even be a lot bigger than this. But suppose it's smaller. I should say that that doesn't make the slightest difference. (In fact in my own calculations, when I first pursued this line of thought, I was out by a factor of nearly $10^{10^{123}}$!) Eventually, a lot of the material in this galaxy will collapse into that black hole in the centre, giving you an entropy per baryon of 10^{31}. All right. What about the very end when the whole universe has collapsed? Well, you can estimate. The black-hole calculations don't directly apply, but you can imagine that there is a larger and larger clumping of the black holes. And, for a collapsed universe, you're going to get an entropy per baryon of something like 10^{43}, giving a total entropy of 10^{123}. That gives us a far larger region for the allowed phase space. If the Creator had aimed at random within this allowed phase space we'd have just seen a mass of congealed black holes (or white holes, their time-reverses). This would not be remotely like the universe we see.

At this point, some people would appeal to the so-called *anthropic principle*. That means: we're here. People are around. They exist and need a decently low level of entropy. So that restricts the possibilities. You've got to think about universes in which people exist. But the trouble with that argument is that you don't need much of the universe to exist in a low entropy state in order to get us. You can get away with just a small pocket of low entropy. The cheapest way of making a roomful of people, or even a world full of people, is by random selection. You start with particles sort of coming together at random — and perhaps ten minutes ago that's what you were — and now suddenly you're reading this book. That's the most likely way, except that it would most probably not be ten minutes but half a minute or even less. The most likely way in which we came into existence is by some such procedure. By a statistical fluctuation, if you like, in an otherwise high entropy universe.

Well I don't believe a word of it; and I hope you don't either. Consistently with our being here, the entropy could be vastly higher than it actually is. If that's the only argument you're using, if you're arguing that the fact that we're here is why the entropy is as low as it is, then the answer to the argument is that you don't need to make the rest of the universe to make us. Given that we're here, the most probable thing outside is something completely different from what we see: namely, lots of massive black holes. When you looked outside your room, you wouldn't expect to see a lot of stars; you'd expect to see black. The anthropic argument won't work here.

So in fact, the Creator must have been constrained in building this universe. How much is the Creator constrained? Well, at least within the small region shown on the fig. 4.5. As I remarked earlier, I've drawn it much too big. By how much have I drawn it too big? Well, bearing in mind that there's a logarithm involved in entropy, it turns out that the ratio of the volume of whole phase space to the small region is $10^{10^{123}}$. You'd have to aim pretty close to hit a point that small. Now I'm not saying that the Creator wasn't that clever. What I'm driving at is that we need a *law* which says that the universe had to be like that. And why not, because, after all, aren't singularities places where ordinary physics gives up? We don't know what's going on in singularities. It may well be that the physics which applies to singularities incorporates such a principle. If so, big bangs are constrained in this kind of a way.

Here I want to postulate something I call the *Weyl curvature hypothesis*. It's no more than a hypothesis, as I've just said, but it works pretty well, I think. My point of view is: all right, you've got singularities in the universe, places where we don't understand the physics because the curvature has gone to infinity. But there's *something* physical that's going on. There must be, because of this very very strong constraint upon the entropy state. We don't understand that physics. But it could well be that that physics is such that the curvature is of a particular kind. Now curvature is described by a thing called the Reimann tensor, and relativists like to split this into two pieces: the Weyl tensor and the Ricci tensor.

The Reimann curvature splits up, naturally, in this form. Now if we take black holes, for instance, what happens, near the end, to the curvature? Well you find that the Ricci part is completely dominated by the Weyl part. The Weyl part is much much bigger than the Ricci part. This is because of distortions that are involved, leading to complications at the singularity. But according to the standard big-bang models (the so-called Friedmann models), quite the opposite is the case in the initial singularities. The Weyl part is much much less than the Ricci part. Of course you can't see these singularities. That's what one deduces, if you like. What one sees, in fact, is the expansion and the black-body radiation, the latter being a relic of the hot big bang.

We don't yet understand the physics of singularities. It's probably a part of *quantum gravity*. But I'm hypothesizing that this will turn out to have, as one of its implications, that in *initial*-type singularities the Weyl curvature is zero. That's a constraint which I'm suggesting this physics imposes. Now this is a time-asymmetric law. I said that the laws of physics that we know and love are time-symmetrical. Well, we don't understand and can't love all the laws of physics, because we don't know them. But these unknown laws, whatever they are, should, I claim, have this implication that the Weyl curvature should be zero, or close to it, in initial singularities.

We're singling out this distinction between initial and final; that has to do with which side of the singularity the part of the light cone that we call the future cone, or the *plus cone*, lies. The plus and minus cones are only different if one believes that there is a time-asymmetry behind them somewhere, in the physics. Well, we know there is. When I say we know there is, it's not utterly and completely clear; but this seems to be a strong implication of the decay of K° particles, which exhibit time-asymmetries in their behaviour. And this appears to be a fact. Of course this may not have anything to do with what I'm saying. It's just that nature seems to have this peculiar property that most of the laws that we see around us are time-symmetric. Yet they're not all *quite* time-symmetric. And the macroscopic behaviour of things seems to be grossly time-asymmetric. I'm saying that if one postulates this Weyl curvature hypothesis, just provisionally, then some of these peculiarities can be explained. The hypothesis tells us, among other things, *how* the entropy was low in the big bang. This is because the Weyl curvature measures the gravitational degrees of freedom. That is basically what it's doing. The Weyl curvature hypothesis says that, for some reason, the *gravitational* part of the entropy was not at all at its maximum at the time of the big bang but was set, essentially, to zero. Somehow the gravitational part of the entropy, the space–time structure part, was not thermalized. Because gravity couples so weakly with the rest of matter, you get away with that for an awfully long time. And only very, very gradually does the gravitational entropy start to build up.

In fact, you might ask: why are we all here, why are all the things around us here? Well, one reason we're all here is that the Sun's out there, and that's a low entropy region. We receive from the Sun a small number of high-energy photons which we busily convert into large numbers of low-energy photons. Correspondingly, the entropy goes up, because there are many more degrees of freedom in the greater number of low-energy photons than in the smaller number of high-energy ones. That's what plants do for us: they sit there and take up these high-energy photons and spew out many more lower-energy photons. That's the way they grow. Then we eat them or we eat animals which eat them and so on. That's how we get around and exist without succumbing to the second law of thermodynamics.

But that raises a further question: why is the Sun there? The Sun is there because it has condensed out of a gas. Previously we had a uniform distribution of matter. If the distribution hadn't previously been uniform we couldn't have ended up with a universe such as we actually see around us. The big bang itself was, in fact, very uniform in structure, as I remarked earlier. Space itself was uniform, with an initially even distribution of gas. And it took a long time before that gas started to clump. It couldn't begin to do so until the expansion had started to die off. Then, as the expansion slowed, you began to get clumping into galaxies and stars and so on. So that all makes sense. But there is no such constraint at the end. The entropy just goes sailing up, and the entropy becomes absolutely huge. You have a very complicated singularity structure. It is nothing like what happens at the beginning.

Well, as I said earlier, some people might consider what I have been saying to be controversial. I'm not sure I understand why. I consider what I've been saying so far to be the uncontroversial part of my thesis. I want to proceed now to the part I admit to being controversial. Where might the Weyl curvature hypothesis come from? I don't think we should just pull it out of a hat, though that's one way of doing it. So let's see if we can do any better. Well, it seems to me that the Weyl curvature hypothesis must have something to do with the, at the moment, non-existent theory of quantum gravity.

That's a rather vacuous statement. But a lot of people are striving towards attaining a theory of quantum gravity, and fine: good idea. But the way in which people try to do this, on the whole, is to take gravity, general relativity, and take quantum mechanics. First of all, they try and do a sort of quantizing of the classical theory, which means applying the rules of quantum mechanics to general relativity and seeing what they get out. Well, I should say that these attempts have been rather dismal in their results. There's a lot of work, a lot of very good work on the subject, but no viable theories seem to come out of it. In fact, the theories that emerge appear to tell you that there's something wrong, that straight general relativity combined with quantum mechanics leads you to something which is *non-renormalizable*, in which totally intractable infinities appear. (Infinities appear in standard quantum-field theory, too. But here there is an *ad hoc* procedure, known as *renormalization*, for dealing with them. Non-renormalizable theories are ones that generate infinities which are not susceptible to being 'tamed' in this way.)

This leads some people to say: well, perhaps we should change general relativity. And perhaps we should. But my own prejudices are the other way around, that if we have this kind of conflict, we ought to do something about quantum mechanics. I'm not saying it's completely the other way around. No doubt one will have to do *something* about general relativity. It certainly can't be totally correct, as it stands. After all, you can't have a classical theory and a quantum theory just sitting there together like that. They have somehow to interact with each other. So general relativity will have to give a little: it will have to have a quantum form. But I believe that quantum mechanics likewise will have to give a little, in fact probably give quite a lot. And one of the things I have against quantum mechanics, and that a lot of people have against quantum mechanics, is that it doesn't make any sense.

Perhaps I should qualify this. It is, of course, an absolutely marvellous theory. It's marvellous for at least two reasons. One reason is that it agrees with all experiments, enabling you to do all sorts of things that couldn't be done before, and you get marvellous explanations and predictions of phenomena coming out. I'm in no position to talk about many of these, because my understanding of them is inadequate. But the ones I do know about are pretty impressive. There's obviously a lot that's right about quantum mechanics in observational terms. The other thing that's marvellous about quantum mechanics is its mathematical elegance. Mathematically, it is a very elegant theory. I should say there's no question about that. So I would certainly not wish to see quantum mechanics replaced, just like that, by something else. And I don't believe that it will be. I think that it has marvellous strengths, both from the theoretical and the experimental side.

Of course when someone starts with this sort of eulogy, you can tell that he wants to be rude about the subject! Well I've already been rude: I've said that quantum mechanics doesn't make sense. But then you see it *doesn't* make any sense. It's all right when you're looking at simple things like single particles or simple systems of particles. But, as I think most of you know, there are things like the Schrödinger's cat paradox and so on which seem to indicate that, when you have a large structure classical physics applies, and something other than the linear rules of quantum mechanics seems to operate. Is the cat in Schrödinger's box, in his thought experiment, really in some linear combination of dead and alive states? Or is it, in reality, *either* dead *or* alive? Well Schrödinger himself certainly thought that it was either dead or alive. And it's not just, as you often read in popular accounts, a question of its being half dead and half alive. According to quantum mechanics, it's in a *complex* linear combination of dead and alive. It could be dead plus alive. It could be dead minus alive. It could be dead plus the square root of minus one times alive. All these things are different and, in principle, distinguishable physically. Now that's what quantum mechanics appears to say and it's difficult to take seriously; because that's probably not what happens. But if it's not what happens physically, then there has, at some stage, to be a modification in quantum mechanics.

The question, then, is what form this modification should take. Before giving you my ideas about this, let me give a more precise statement of the Weyl curvature hypothesis. It seems to be possible to divide all singularities into two mutually exclusive and jointly exhaustive classes: past singularities (such as the big bang) and future singularities (such as the big crunch or the singularities contained in black holes). What the hypothesis then says is: *the Weyl curvature tends to zero at all past singularities, as the singularity is approached from future directions.* Two things not so far mentioned follow from this. First, the principle rules out the existence of so-called *white holes*. These are the hypothetical time-reversed counterparts of black holes: a white hole would spew out matter just as black holes suck matter in. Just as you can't get out of a black hole, so it would be impossible, without exceeding the local light velocity, to get into a white hole. There have been a number of intriguing speculations concerning white holes. One is that we are actually living in a white hole. That is observationally wrong, since we have seen that the big bang singularity is actually not at all like that in a black hole. Another speculation is that the matter sucked in by a black hole at one point in the universe is emitted by a white hole at some other point, perhaps thousands of light years distant. Well, I'm sorry to be a spoilsport. But the Weyl curvature hypothesis unambiguously rules this out too.

The second and related thing that follows is that we get a problem when we consider regions of phase space that are supposed to contain black holes. Once we rule out white holes, it turns out that there are more paths going into such a region of phase space than there are going out. And this violates something known as *Liouville's theorem*. This is because lots of widely divergent conditions at one time can, if they lead to gravitational collapse into a black-hole state, result in identical states at a later time. Black holes, as the saying goes, 'have no hair': they retain no trace of their precise origins. What I want to suggest is that a proper quantum theory of gravity, once we have such a thing, will allow for actual physical collapse of the wave function; and that this physical collapse compensates for what happens elsewhere in the phase space, when a black hole forms. Just as there are more paths into than out of a point in phase space corresponding to a black-hole state, so there would be more paths out of than into a point of phase space corresponding to a physical collapse of the wave function. When we have both phenomena, this allows for an overall balance within the total phase space. For the balance to be achieved, we require that wave function collapse be a *gravitational* phenomenon. An important point is that quantum mechanics tells us that identical wave functions may collapse into quite different states. For example, in the case of Schrödinger's cat, half-dead plus half-alive may collapse either into dead or into alive. Indeed, the whole point of what I'm now saying is that it *must* do so, in this sort of case. Big systems cannot, if I am right, exist in linear combinations — *superpositions* as they are called — of macroscopic states.

It remains to consider under what conditions, precisely, the wave function should be expected to collapse. Well, from a certain point of view, a physical collapse of the wave function ought to be associated with a fall in entropy — if we ignore gravity, that is. (This is a 'commonsense' entropy and not 'quantum-mechanical entropy'.) That would explain why we don't get spontaneous collapse in very small systems: for them gravitation is of negligible strength. But, as we consider progressively larger systems, gravitation becomes progressively more significant. Moreover, gravitational entropy, as we have seen, rises as matter becomes more concentrated. So it's a question of balancing these two effects. My suggestion is that the wave function collapses — or at any rate is physically *allowed* to collapse — when the system is sufficiently massive for such collapse to result in a net gain of entropy. In other words, the wave function only collapses when it can do so consistently with entropy going up, or at least not falling. To calculate when this would happen, one needs a general way of measuring gravitational entropy. The Hawking—Bekenstein formula is just one example of a formula that assigns an entropy to a space—time geometry; but we want to be able to do this quite generally. There is a particular formula that I have been working with, which is a generalization of the Hawking—Bekenstein formula; and I have tried applying this to the problem of droplets condensing in a cloud chamber. What I end up with is figures around a millimetre for the diameter at which a physical collapse of the wave function should be expected to occur. This is perhaps a bit on the large side, but not ridiculously so. (It would suggest that anything from fleas on down can exist in linear superpositions of states, but cats certainly can't.)

I must emphasize that these ideas are highly speculative and not at all fully worked out. But I find them suggestive. There is something deeply unsatisfactory about the present conventional formulation of quantum mechanics, incorporating as it does two quite distinct modes of evolution: one completely deterministic, in accordance with the Schrödinger equation, the other a probabilistic collapse. And it is a great weakness of the conventional theory that one is not told when one form of evolution is supposed to give way to the other, beyond the fact that it must always take place sometime prior to an observation being made. (Some, like Wigner, have even suggested that it is consciousness that collapses the wave function.) The foregoing line of thought holds out some promise of our being able to synthesize these two elements into a single, unified theory. Obviously, if I am right, the Schrödinger equation will have to be modified in some way. And just how I am not sure. It is even possible that we may end up restoring determinism in quantum mechanics, which would have pleased Einstein. Personally though, I doubt that it would be quite like that; and it is not, in any case, the statistical character of contemporary quantum mechanics that creates the problem. What we are after, as I say, is a theory that (a) accommodates gravity and (b) makes sense. Some people seem content to regard quantum mechanics simply as a useful computational tool, not worrying about its interpretation. But, as a philosophical realist, that doesn't satisfy me. I cannot believe that it is impossible to make sense of the physical world. I think that it is the task of physics to do precisely that.

POSTSCRIPT

A while after giving the lecture on which this article was based, I somewhat shifted my view as to the criterion determining physical wave-function collapse (partly as a result of objections by R. M. Wald). I do not now think that the criterion should be phrased in terms of a concept of 'gravitational entropy' but, rather, the more clear-cut 'graviton number'. On my *present* view, wave-function collapse takes place when the difference between the gravitational fields of the states under superposition amounts to at least one graviton's worth. Combining this reinterpretation with a refined calculation by A. Ashtekar, we can reduce the mass of the droplet by a factor of about a million — a considerable improvement.

FURTHER READING

Landsberg, P. T., *The Enigma of Time* (Adam Hilger, Bristol, 1982). This book contains an excerpt which is reasonably accessible to the non-specialist from R. Penrose's paper 'Singularities and time asymmetry' which appears in full in: Hawking, S. W. and Israel, W. (eds) *General Relativity — An Einstein Centenary Survey* (Cambridge University Press, Cambridge, 1979).

Layzer, David, 'The asymmetry of time', *Scientific American*, Dec. 1975. This article discusses further the conflict raised by Dr Penrose between thermal equilibrium and increasing entropy and defends the viewpoint that Dr Penrose questions.

Further details of the Hawking—Bekenstein formula may be found in the following technical articles: Hawking, S. W. *Communications in*

Mathematics and Physics, 43 (1975) p. 199 and Bekenstein, J. D., *Physical Review*, D7 (1973), p. 2333; 9 (1974), 3292.

A proposal to introduce consciousness directly into quantum mechanics is given by Eugene Wigner in *The Scientist Speculates*, ed. I. J. Good (Heinemann, London, 1962)

5

Time and Cosmology

MICHAEL SHALLIS

An understanding of the nature of time has always been intimately connected with mankind's understanding of the universe as a whole, of the cosmos, and hence with cosmology. However, I would suggest that we are unlikely to discover some ultimate truth about the nature of time in any philosophic or scientific enquiry; that the most we can hope for is to understand how a particular culture thinks about the nature of time, perceives and describes it and how that culture's perception of time reflects and influences its cosmology. In the context of our own culture and the scientific disciplines that are dominant within it (in the case of this chapter, physics and astronomy) we can try to understand how the prevalent assumptions about time in our culture affect and reflect our cosmology and how these assumptions are extended and explored in our science.

One of the ways in which a culture's cosmology is expressed is through the sort of technology it produces. As we are concerned with time, we might start by examining the ways in which different societies measured, or rather for most cultures indicated, time. Right away we find that there is an inevitable connection between the indication of passing time and astronomy. The fundamental perception of all mankind is the distinction between day and night, of the phases of the moon and the passage of the planets through the realm of the fixed stars. Time, in this context, is organic, cyclic and sacred. We find its mysteries embodied in the great stone circles, such as Stonehenge. The important thing for societies with this basic and natural perception is not the passing of discrete moments but the re-enactment of the phases of the cycle. To what the anthropologists call 'primitive' people (but it is true of all sacred and religious cultures), religious ritual in the repetition of a primordial or archetypal action, not just repeated at another time but through the ritual actually repeating that original moment, in this way past and present link totally so that time ceases to have meaning in the sense of a historical continuum.[1] It is quite a different sort of time from that of our own cosmology.

Time flows and many timekeeping devices demonstrate this flow. The hour glass is actually a very good clock in that it models the flow of time and the unidirectionality of that flow. Indeed, one of the reasons that we think of time as only flowing in one direction is the very same as the reason that sand in an hour glass does not spontaneously return to the upper chamber. (But that is a topic which has to do with the concept of entropy dealt with elsewhere in this volume: see chapters 3 and 4.) The hour glass, of course, while it indicates time's flow, does not measure time beyond marking off a single, rather large interval. In contemporary culture we have become obsessed with timekeeping and clock-watching, down to the level of micro- or even nanoseconds, so it is salutary to be reminded that perception of duration is also something culturally determined. In India, before modern times, the shortest time interval that could be meaningfully discussed, because there were no words for anything shorter, was the time it took to boil rice (about 13 minutes). Just for a moment imagine what that might mean to the way you thought about the world!

The sundial perhaps most of all reflects the direct connection between time, or timekeeping, and astronomy. The shadow cast by the moving sun links our sense of time immediately to astronomical phenomena and examples of sundials range from the upright, medieval column sundial to modern, sophisticated and highly accurate ones, such as that found in the gardens of the Royal Greenwich Observatory at Hurstmonceaux. It is only very recently that our standard of time, the definition of the second, has changed from being based on a particular fraction of a year (the Earth's passage around the Sun), to being defined in terms of atomic oscillations. The atomic clock depends not on the movement of the heavens but on the less visible physics of the electromagnetic properties of matter. In this sense we can say that we have decoupled our perception of time from our cosmology, or at least from our perception of the universe.

The history of our culture has, from one perspective, been a history of the progressive pinning down of time, of making ever more accurate clocks. The graph in fig. 5.1 shows the exponential increase in accuracy of clocks from the Middle Ages to the middle of the twentieth century. It has been correlated (not entirely spuriously) with many other patterns of growth, including the frequency and ferocity of battles and warfare in Europe.[2] What it tells us is that our perception of time is essentially linear, that we think of time as a continuum that can, in principle, be subdivided into ever finer gradations, as if it were possible to pin things down to exact nanoseconds. Time, in our world, is that quantity measured by a clock. This process has its origins in the great monasteries of the

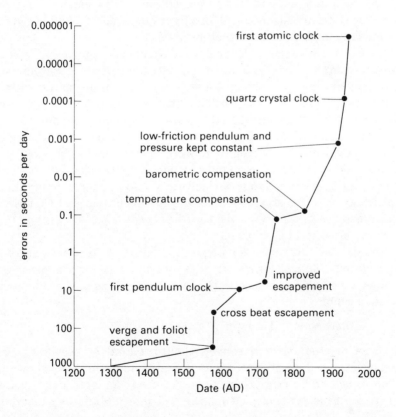

Fig. 5.1 *Chart showing the rise in accuracy of clocks through the centuries, and major improvements to mechanisms (based on the chart given in J. -F. Phipps'* Time and the Bomb[2])

Middle Ages, where the bell and then the clock regulated the time for prayers and liturgical office. However, the medieval clocks, whilst propagating our own perception of time, still embodied a medieval and pre-Copernican cosmology. For example, the famous clock in Wells Cathedral is both astronomical and astrological. It stems from a more ancient world view, a cosmology quite alien to our own. It derives from a cosmology that conceives the universe as a hierarchy, stretching from the earth up to heaven; and all art, science, politics, theology is incorporated into its scheme. It contains time in its description at the level at which time appears to us, as observers, in the passage of the stars and planets, just as time in the form of the passage of the sun appears in, for example, Sumerian cosmology, or even in a description of the Copernican cosmos. This is inevitable, since our individual and human perception of time cannot really be separated from astronomical observation, even if contemporary cosmology appears to force that separation.

Just as we have decoupled our standard of time from astronomical observation, so too have we decoupled our cosmology from our religious and political thought and, seemingly, from the rest of our culture. At least, we may believe that we have done so, in the abstract and mathematical modelling of the universe that we refer to as the science of cosmology. I would suggest, however, that our cosmology, like all cosmologies, reflects what we are like as a culture, and would even add the comment that to an outside observer (if there can be such a thing) our culture would appear unintegrated, rationalized and abstract — a culture in which man and God have been displaced and where even time has been dispensed with by the convenient trick of spatializing it.

In our contemporary cosmology, time figures as just one axis of a spatialized reference frame. It has become a 'fourth dimension', rendered commensurable with the three dimensions of space by way of the formula

distance = speed of light × time.

In this manner, time is converted into a kind of pseudo-spatial dimension, and what emerges is a static, four-dimensional world view. Space and time give way to *space—time*. A single, frozen point of space—time can be depicted in the well-known Minkowski diagram (fig. 5.2). This shows how the limiting value of the speed of light, defining a cone of light rays that converge on the here and now, divides the universe, as the observer here and now sees it, into

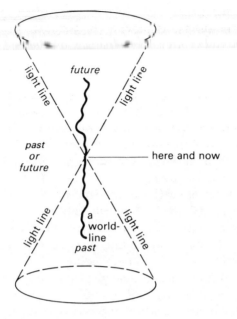

Fig. 5.2 *A Minkowski space—time diagram, showing the past and future regions of space which are accessible to an observer at 'here and now' (based on a diagram from Michael Shallis' On Time[3])*

accessible and inaccessible regions. In the presence of matter and energy space—time becomes curved, distorted, perhaps even forming tunnels, loops and other bizarre topological structures.

Our cosmology, therefore, is an abstract and mathematical one. We no longer envisage the universe in terms of absolute space, a vast three-dimensional matrix in which things happen at specific moments according to some absolute and universal clock, but as one in which time is relative, passes by at variable rates for different observers and yet, seemingly, can be thought of as ticking by uniformly for the universe as a whole. Or can it?

Dr Sciama has, in chapter 2, explored the consequences, for time, of Einstein's special and general theories of relativity. According to Einstein, the rate at which time passes for a given observer — the elapsed time which will be registered by a properly adjusted clock that shares the observer's state of motion — will depend both on his velocity and the local gravitational potential. Time, if we follow

Einstein in equating time with the behaviour of an ideal clock, will run more slowly on faster-moving bodies and in more intense gravitational fields. Both these effects have been experimentally verified using identically constructed and initially synchronized atomic clocks: in the one case, by taking one such clock on a round-the-world trip in a jet aircraft, and then comparing it with another that had remained on the ground; in the other case, by comparing two such clocks placed at the top and bottom of a tall building, where the gravitational potentials would be slightly different.

Is it possible then, in the light of such phenomena, to think of the universe as a whole as having a definite age? If, for example, time runs more slowly on the surface of the Sun than it does on the surface of the earth, can we still think of the Sun and the Earth as having the same age? Here we must make a distinction between *local time* and *coordinate time*. (What I am calling *local time* is what physicists often refer to as *proper time*; I am avoiding that term here since, as we shall see in a moment, cosmologists use *proper time* to mean something different.) Every localized object in the universe — and that will include human (or other) observers — can be thought of as having a local time, that passes at a rate which is determined by its state of motion and the gravitational potentials it encounters. An ideal clock that moves with an object is registering its local time, in this sense. If we think in terms of that object's *world-line*, that is to say, the path through space—time that corresponds to the history of that object, then what a clock that travels with that object is really doing is measuring off the space—time length of its world-line; the elapsed local time, between two space—time points on an object's world-line will simply be the space—time length of that segment of the object's world-line which links these two points.

It is also possible, however, to take a single clock as standard, taking it to define a universal time coordinate, and to relativize everything to it. This *coordinate time* may then be thought of as ticking away uniformly for all objects, regardless of their state of motion or the local gravitational potentials; and the elapsed coordinate time will, in general, diverge from the local time. Of course, the choice of a coordinate time is, to a considerable extent, arbitrary: in principle, one could take any clock as one's standard. But in a cosmological context it is natural to take as standard a clock whose motion is *typical* or *representative* of the motion of matter in general — one which simply 'rides along', so to speak, with the overall expansion of the universe. This is what cosmologists refer to as *proper time*. I shall be coming back to it later.

For a moment let us return to the horizons and see how modern astronomy has fundamentally changed the model of the universe. We shall then be better able to address such questions as how old the universe is.

The night sky is filled with stars, especially in the band of light we call the Milky Way. At the turn of this century, indeed up to the late 1920s, it was generally believed that the Milky Way, our galaxy, was the universe. In the 1920s a great debate took place as to whether the spiral nebulae, like the Andromeda nebula, were other 'island universes' or were simply local objects in our own system. We now all know that the universe is filled with spiral galaxies, each containing around 100,000 million stars, with countless elliptical galaxies, ranging in size from hundreds of millions of stars to 10^{12} stars. There are also irregular galaxies, highly energetic galaxies, galaxies that look rather uninteresting, but which emit radio wavelength radiation of great magnitude from huge regions of space around their optical counterparts and the QSOs, the quasi-stellar objects, which many astronomers believe are the most distant astronomical objects we observe.

More importantly, galaxies are observed in groups, like Stephan's Quintet, and in clusters of sometimes thousands of galaxies. We now know that the universe contains hundreds of thousands of millions of galaxies, yet it is well to remember that this perception of the cosmos is barely 30 years old.

In the 1920s Edwin Hubble, an American astronomer, working on the controversy about island universes, took spectra of galaxies using the big telescopes at Mount Wilson and Palomar. His discovery led to the second fundamental observation on which modern cosmology is based. (The first being that the sky is dark at night, which led Olbers to pose his paradox in the nineteenth century.[4]) Hubble observed all galaxies to have a red shift. Interpreted as the Doppler effect, the red shift indicated that all galaxies were moving away from our own, and furthermore that the further away a galaxy was the faster it was moving away from us. Hubble was able to show that there is a linear relationship between the velocity of recession of a galaxy, measured by its red shift, and its distance. This relationship is known as the Hubble law:

$$cz = Hd,$$

where c is the speed of light, z is the red shift — measured as a relative change in wavelength — H is Hubble's constant and d is the distance to the galaxy. Hubble's observations and Hubble's law tell us that the universe is expanding.

Most modern cosmology is concerned with questions related to this expansion, to its origin and its likely future, but before discussing those questions I would like to make a few comments about the difficulties inherent in cosmological speculation. There is a wide discrepancy between what can be imagined in terms of mathematical and physical theory and modelling and what can be determined observationally.

For example, we know from Hubble's observations that the universe is expanding, that typical scale lengths in the universe are increasing with time and that the expansion rate was greater in the past. This is known theoretically. There was a time when scale lengths were zero and the expansion rate was infinite — the so-called big-bang origin of the universe, but when was that moment? We cannot even determine the present expansion rate, Hubble's constant, to better than a factor 2, let alone the more interesting things like the deceleration parameter, the factor by which that expansion rate is slowing down. If it has a value of less than 0.5 (in the parameters that it is worked out in) then the universe is open and will expand forever. If it is greater than 0.5 then the universe is closed and will recollapse in an inverse big bang — a big crunch. However, determining the deceleration rate depends on making observations at different epochs, by looking back in time by looking further and further away from our own galaxy in order to see how the expansion rate varies. The trouble is that we cannot see far enough away, far enough back in time, to see deviations from the linearity of Hubble's law. If we could there would still be problems, because we would have to correct the measured luminosity of very distant galaxies due to restrictions of telescope aperture and wavelength bandpass of the detection system. There would also be an unknown correction to be made for evolutionary changes to the luminosity of galaxies. We simply do not know how the brightness of galaxies varies with time, although we do know stars get dimmer with age. Galaxies, however, could brighten with time if more stars are evolving within them.

The deceleration of the expansion rate also depends on the amount of material in the universe. If we knew how dense the universe was we could calculate the deceleration. The problem presented with this is known as the missing mass problem. Huge discrepancies arise in the ratio of mass to luminosity when looking at the dynamics of clusters of galaxies in comparison to the ratio for individual galaxies. There does not seem to be enough material to keep the clusters together, although they are clearly gravitationally bound. This problem translates simply into the fact that we do not know the mean density of the universe.

The Hubble relationship is determined from measurements of luminosity and not distance and there is some controversy, especially relating to QSOs, about converting red shifts to distances. It could be that some part of the observed red shift is not due to the Doppler effect which only serves to confuse the issue further. In any case the luminosity/distance calibration is poorly founded, even within our own galaxy, and so could be doubly misleading if and when deviations from linearity in a non-Euclidean universe show up. The suitability of galaxies as standard candles is not fully justified and galaxies may not necessarily have well ordered and stable luminosities, all of which leaves us knowing the deceleration parameter to roughly 0.5 ± 1!

The fact that observational evidence is hard to come by need not deter us, however, from pursuing this matter. If we look at a space—time diagram that shows the world-lines for some galaxies (see fig. 5.3), there are four things to notice. First of all we see that at successively later times the galaxies are all further away from each other, and, as depicted here, because of the change in expansion rate, the world-lines are curved. The galaxies are depicted as accelerating away from each other or travelling through curved space. That implies that they will be experiencing time differently, that their clocks will be running at different rates and not linearly as the vertical time scale implies. That is really a limitation of this particular form of representation and can be overcome by taking lines perpendicular to each world-line and joining them up as shown in fig. 5.3. The second thing to say is that a set of such lines would then mark the same epoch in the history of all galaxies and would therefore define a universal time that would, to all extents and purposes, be very similar to Newton's idea of absolute time. In practice it would not be possible to make the measurements necessary to set up such a universal time system, but in theory it enables us to talk about universal time.

The third point here concerns the fact that as world-lines curve away from each other so their relative velocities can approach and indeed exceed the speed of light. This is indicated by the right-hand

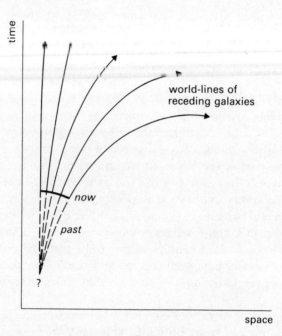

Fig. 5.3 *A space—time diagram showing the world-lines of receding galaxies in an expanding universe (based on a diagram from Michael Shallis' On Time[3])*

line curving until it is perpendicular to our world-line. Now, it is commonly known that nothing can reach the speed of light, but that is not a total restriction on relative motion. The only thing that can reach the speed of light is . . . nothing; no material body can be accelerated to the speed of light. However, in the case considered here no material thing is being accelerated to the speed of light. A galaxy, of course, is a material thing, but these galaxies are not being accelerated away from each other at all — it merely looks like that. What is actually happening is that the space between the galaxies is expanding. It is the universe, the space—time matrix, that is expanding. The galaxies are only moving away from each other because they are embedded in that expanding space. There is no law of physics that prevents there being so much space between galaxies that their relative velocities exceed the speed of light. All this with space expanding at a fairly modest rate — about 60 kilometres per second for every Megaparsec. This means that any two galaxies 5000 Megaparsecs away from each other will be travelling with relative recessional velocities of the speed of light. They will have infinite red shifts with respect to each other and will be invisible to each other. We can even postulate that that distance, 5000 Megaparsecs, is roughly the diameter of the visible universe from any point within it. It is almost as if we live in a super black hole, whose event horizon is 5000 Megaparsecs away, which is almost a pre-Copernican notion. (Incidentally, a Megaparsec is a million parsecs and one parsec is about three light years or 60 million, million miles.)

As an aside here, let me just comment that the speed of light is an energy barrier; you cannot provide any particle with enough energy to accelerate it to the speed of light, although the speed of light is a finite velocity. On space—time diagrams it is often conventionally drawn as a 45° line, separating space—time into two regions of which only one is accessible to us, and of which the Minkowski diagram in fig. 5.2 is an example. As time slows down the closer you approach the speed of light, so one can deduce that for a photon, a particle of light, time does not exist. To that extent a photon can travel right round the universe in no time at all, because in its reference frame there is no time.

The last of the four points concerning fig. 5.3 is that we can extrapolate back from now and suggest that the galaxies merged some time ago, before which was the big bang. And when was that? I am not going to recount the history of the universe back to the big bang but merely remind readers that the generally accepted cosmology of the origin of the universe runs broadly like this.

The universe 'begins' (and I use that word cautiously because we have not and are incapable of defining what we mean by 'begins' in any physical sense) as an infinitely hot, condensed explosion of energy. It is believed that at a very early epoch in the history of the universe (around 10^{-39} seconds) the four forces of nature — the gravitational, electromagnetic, strong and weak nuclear forces — were united as one single cosmic force and became decoupled as the whole universe expanded and cooled. In the very early universe vacuum fluctuations in the intense energy matrix led to particle production and when time had passed by and the universe was around 10^{-6} seconds old familiar particles like protons, electrons, neutrons and neutrinos were created. Matter condensed out of radiation as the universe cooled and expanded and its energy density therefore diminished. At about one second into the life of the universe helium and deuterium nucleii were formed but the temperature of the expanding universe was dropping rapidly so that after about 1000 seconds all that was left was a radiation dominated fog of ionized plasma, something like the interior of the Sun. It is opaque, expanding, cooling. After about a million years the temperature of the universe had dropped to around 3000° and suddenly electrons recombined with nucleii to form stable atoms and the universe became transparent and matter dominated, just as it is today, except that it is now very much cooler, only 3° absolute, and filled with stars and galaxies.

This is a very brief account of what is a very detailed scenario, but it is one that is commonly known and which can be read about in more detail in several popular books.[5,6] What is less familiar is, perhaps, the timetable for the future of the universe. In the first case suppose the universe is closed and will recontract. Once the universe reaches its turning point after about another 10^9 years (remember the universe is roughly that age now) the clusters will start to merge and about 10^8 years later the galaxies within the clusters will start to collapse into each other. A million years later individual stars, pulled into this collapse, will be accelerated to speeds approaching the speed of light (but not reaching it) and a hundred thousand years later the night sky would be distinctly hot and bright. A thousand years later it would be too hot for stars to remain as viable entities and about one year later the universe would return to the state of a very hot, ionized plasma and then rapidly into bizarre states as the universe approached a singularity. Space and time would be highly distorted if space and time remained meaningful concepts.

The future of an ever-expanding universe is perhaps even more interesting. The age of the universe is now about 5×10^9 years; when it has reached 10^{14} years all the stars will have run out of fuel, they will have burned up all the hydrogen that is available to them and will collapse to the white-dwarf state. At around 10^{17} years any planets there are will have left their parent star (the Sun in our case) due to random encounters with other stars. When the universe is ten times older still galaxies will have lost most of their component stars, for the same reason. Any remaining stars will have coalesced to form supermassive black holes. From about 10^{20} years cold stars will start heating up because their constituent protons will be decaying at a rate sufficient to release energy in significant amounts. That is assuming that protons decay, as many physicists now believe. By around 10^{30} years and certainly by 10^{40} years, most matter will have decayed and the universe will consist mostly of electrons, positrons, neutrinos, photons and black holes. The temperature will be around $10^{???}$ degrees absolute and the radius of the universe will be about 10^{20} of its present radius. After about 10^{60} years the black holes will begin to evaporate significantly, due to quantum effects, and they will more or less all have evaporated after 10^{100} years. Incidentally at around 10^{70} years electron/positron pairs will form something a little like atoms as the particle pairs revolve around each other and slowly spiral in on each other to their eventual mutual annihilation. These 'atoms' will each be larger than the present size of the universe!

A little later, at around 10^{1600} years, white dwarfs will decay to become neutron stars — if proton decay does not actually occur — and, still assuming no proton decay, after $10^{10^{76}}$ years (ten to the power of ten to the power of 76) the neutron stars will also have decayed, through tunnelling effects, into black holes which themselves rapidly evaporate leaving a universe filled only with radiation. It is interesting to note that in this scenario we find a universe that began with pure radiation, out of which matter was formed, and end up with all the matter finally returning to its initial state and the universe is once again pure radiation, even if it takes $10^{10^{76}}$ years.

This is, of course, a very remarkable projection and contains a great deal of serious theoretical physics. The question to be addressed, however, is what is actually meant by time when discussing the past or future of the universe. Surely, time in a highly condensed universe will be flowing at a much slower rate than in a vast, empty, low-density universe. One trouble is that the universe is too uniform now for us to be aware of changes in the rate of clocks due to the changing density of the universe. What I want to ask is how long, really, were those first three minutes?

This is not a facetious question because it raises a very important point about time and cosmology. It raises the question of what time scales are appropriate, a matter raised earlier in this chapter. When we talk about the first 10^{-33} of a second, or the first three minutes, or the first million years of the universe we have in mind what we earlier referred to as *proper time*. Now, proper time is time measured by an ideal clock, stationary and in the same gravitational field as an ideal and hypothetical observer. It is a form of Newton's absolute time, except it is confined locally, although it refers to the universe as a whole. Proper time measuring those first three minutes would indicate that those minutes were the same sorts of minutes we are experiencing now. But such a definition of time is not only hypothetical and abstract, it is also misleading in many ways. For a start it ignores changing clock rates as gravitational field strengths change and it does not relate to the physical properties of the universe generally.

If we decided that some physical event defined a unit of time, for example the mean free time between particle interactions, chosen in some relevant way, then we would find the early universe was very full of events and the future universe very empty of them. That is, *physical time* ran a lot faster in the early universe and is continually slowing down. On such a scale the universe would effectively be infinitely old. To handle such ideas physicists such as York and Misner have developed technical time scales that model different epochs in different ways. *York time*, for example, starts at minus infinity, the big-bang origin of the universe, and proceeds to time zero in the distant future. Using such a scale the universe is infinitely old but has a finite future, unless it all recollapses, which of course will take an infinitely long time. Using proper time, in contrast, the universe is scaled as having a finite past but an infinite future (unless it recollapses, when its end will be in a finite time).

Universal time, in the sense of mapping out the epochs in galaxy world-lines, is a form of physical time and leads to infinities near the big bang, but all these ideas are concerned, once again, with abstractions. We do not learn anything about the nature of time from discussing them, only about the way we choose, for convenience, to scale time. What about clocks themselves, what sort of time do they measure?

There are basically two sorts of clock, gravitational clocks, such as hour glasses and pendulum clocks (and I include spring-driven watches, electric clocks and so on in this class, because the spring, battery or whatever is only a device to provide the equivalent of gravity for driving the device), and atomic clocks, which depend not on the gravitational force (although they are affected by gravity) but on the electromagnetic force. An alternative way of distinguishing between these two forms of clocks is to base their timekeeping on the gravitational mechanics of the solar system or on the properties of atoms. The ratio between the gravitational and electromagnetic forces is 10^{40}, but if the forces were initially unified then the scale differences between these forces should have been less in the past. It has been suggested that the gravitational constant (G) has changed with time, mostly in the very early stages of the history of the universe. If G has changed with time then gravitational clocks have run at different rates, fundamentally, from atomic clocks. If they both measure some form of proper time it would still be necessary in referring to the age of the universe to specify which sort of clock was used to measure that age with. In other words, we may not be able to specify the universe as having a unique age at all.

As a final example of modern cosmological speculation concerned with space and time I want to mention the tachyon, a hypothetical particle that only travels faster than light. Tachyons are simply mirror images of normal particles, with the reflection being in the speed of light line on a space—time diagram. Tachyons can only travel faster than the speed of light, so they inhabit the half of the diagram that is forbidden to us. As the tachyon gains energy it slows down and as it looses energy it speeds up, the inverse of what happens to particles in our half of the diagram. You cannot give it enough energy for it to slow down to the speed of light, just as you cannot supply enough energy to a normal particle to accelerate it up to the speed of light. As you take energy away from a tachyon it speeds up until you cannot take any more energy away from it. At such a point it has reached the equivalent of its 'rest mass', in this case its 'infinite velocity mass'. After all it cannot go faster than an infinite speed, just as a normal particle cannot go slower than zero speed.

Now consider a tachyon created in the big bang. It travels out into the universe faster than the speed of light and looses energy as the universe expands, so it speeds up (fig. 5.4). Eventually it reaches an infinite speed and has travelled a distance, d, the tachyonic radius of the universe. However, the universe still expands so the tachyon has to loose more energy and can only do so by going backwards in time. (Actually it was going backwards in time before, because if you travel faster than the speed of light you automatically go backwards in time, like overtaking a slower train.) The result is that the tachyon ends up back at its moment of creation, but a distance $2d$ away.

Of course, you can regard this situation as a picture of a tachyon/anti-tachyon pair emerging from the vacuum at vast distances away from each other and mutually annihilating on collision. Otherwise you have to admit that the tachyon is at two places, widely separated, at the same time. It is always at two places at once, except where it turns round. To the tachyon it is space that flows in one direction relentlessly, whilst time is a dimension easier to move about in. It is a time-like world, whereas ours is a space-like world. Theoretical speculation can lead into quite unimagined and unimaginable realms.

We have covered a lot of ground in this somewhat compressed

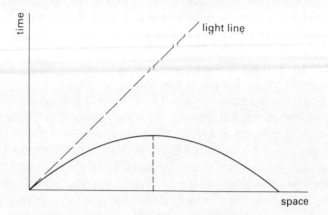

Fig. 5.4 *A space—time diagram for the hypothetical particle, the tachyon, showing how its path reverses and interconnects places at great distance from each other at the same time (based on a diagram from Michael Shallis' On Time[3])*

overview of current cosmology. What we have ended up with is a view of at least a part of our own culture's world view. It is a cosmology that is highly technical, reflecting a society that has become almost obsessive about technique, both in its physics and in other aspects of culture; it is mostly abstract, in many ways it is quite arbitrary, as theories can be produced to describe or elaborate on almost anything, just as it is also constantly changing (this year's cosmology may be quite different from last year's). It is a cosmology which I think tells us little, if anything, about the nature of time, except perhaps that time is a lot stranger than we might think. I do think it tells us a great deal about ourselves, although it is not normally regarded in that light. It tells us of our obsession with technique (we will solve the puzzles that remain if we develop better instruments), about our abstractedness from 'normal' human experience, about our inconstancy, our insatiable need for and acceptance of perpetual change, our lack of a base, if you like.

Our cosmology also tells us about the way we perceive our relationship with the cosmos and with nature. That relationship, our cosmology tells us, has ceased to be grounded in the organic and cyclic world of nature in which we live our daily lives and in which time is more explicitly, more immediately present. Cosmologies have always incorporated a society's myths into their presentation and explanation of the world. There is no reason to suppose our cosmology is any different. It too is our myth, but it lacks the symbolic richness, the human ideals, of past myths. To that extent it tells us a lot about ourselves and the culture we have built up over the past 300 years or so. However, the theme of this chapter involves cosmology with time and if I had to encapsulate what we have learned about time from our cosmology it would be to say that we appear to have abstracted time, to have lost it or simply that we have passed time by.

FURTHER READING

1   Campbell, Joseph, *The Masks of God* (Penguin, Harmondsworth, 1976).
2   Phipps, J. -F., *Time and the Bomb* (Pica Press, Oxford, 1982).
3   Shallis, Michael, *On Time* (Penguin, Harmondsworth, 1983).
4   Narlikar, Jayant, *The Structure of the Universe* (Oxford University Press, Oxford, 1977).
5   Barrow, J. D. and Silk, J., *The Left Hand of Creation* (Heinemann, London, 1983).
6   Weinberg, Stephen, *The First Three Minutes* (Fontana, London, 1978).

6

Time and Dispersal: The Second Law

P. W. ATKINS

Time is geometry. Change is a consequence of the purposeless exploration of this geometry. Our consciousness of the passage of time is the chemical imprint on our brains of this purposeless, irreversible journey. The imprint, the memory, and the experience are chemical.

That is the span, expressed somewhat grandiosely, of this chapter, which is essentially an examination of the experience of the passage of time at the level of the atoms and molecules that participate in the events in our brains. It travels from the mechanism of extremely simple events to the principles behind the processes that contribute to our consciousness of the passage of time. This does not mean that it travels far into the elucidation of consciousness, for it is intended to lay only the groundwork of the arguments involved, to identify the spring of the irreversibility of the world, and to show that the processes involved in our perception of time are in principle no more complicated than our perception of any kind of event.

This chapter explores the origin and consequences of a minute fragment of the diagram on p. 81 of the variation of entropy (a concept elucidated later) with time (fig. 6.1). There are only three significant regions in the diagram: region A, when the universe began, region B, one possibility for the end of time, and region C, which is about now. Region A is described in the following chapter; region B is of no immediate concern; hence neither region is discussed here. This chapter concentrates on region C, which, although austerely expressed in the illustration, actually corresponds to life's

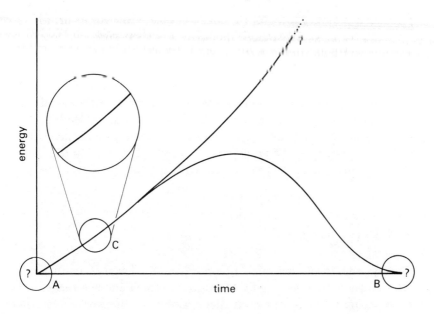

Fig. 6.1

rich pageant and all the glorious events that surround us in the world.

Life's rich pageant is summarized slightly less austerely by the Laws of Thermodynamics, of which there are four. The first to be discovered was the Second, and the second was the First. The Third is possibly not a law of thermodynamics at all, and the Zeroth, discovered fourth, is an afterthought. That is all there is that is complicated about thermodynamics; the application of the laws is very much simpler.

The enlargement of region C can be taken as representing a period of any duration. It can be the time-scale of our lives, a day, a minute, a microsecond, a million years, a thousand million years, or a picosecond. The time at which the region occurs can also be long before the dinosaurs, or even before the emergence of the Earth, it can be this instant when consciousness is triumphant (but perhaps endangered), or it can be far into the future when all conscious life may have ceased.

The principal question addressed by this chapter about region C, is how it allows for the emergence of constructive events. In particular, how it allows for the construction of cathedrals or, what is very much the same thing, of thoughts and memories. Our thoughts and perceptions leave a trail of chemical footsteps on the chemical composition of our brain, and the more or less permanent impact of those footsteps on the future responses of our brains helps us to identify and experience the passage of time. The theme of this chapter is to discern the processes that ultimately lead to this consciousness of the passage of time, and to show how our perception of time is rooted in the simplicity of events that are committed by their intrinsic nature to purposeless drift into the future.

THE SECOND LAW

This chapter deals with the Second Law of Thermodynamics, and aims to show that, especially through its atomic interpretation, it is an especially clear window on the workings of the world. No other law has contributed so much to the liberation of the human spirit, and hence to the improvement of the human condition, and so it is a particular cause for regret that its presentation so often leaves the impression that it is both difficult and dull. Even the people who first formulated it have changed their names, as though through a sense of shame!

Rudolph Gottlieb proposed a version of the law, and then quickly changed his name. He affected a classical disposition, and called himself Clausius, the name by which he is always known today. The *Clausius statement* of the Second Law is as follows:

Heat does not pass spontaneously from cold to hot.

The second statement of the Second Law is due to William Thompson, who also quickly changed his name, and we now know him as Lord Kelvin. Kelvin's statement of the law is:

Heat cannot be converted completely to work
without another change elsewhere.

Both Clausius and Kelvin focussed their attention on the heat engine, that extraordinary device which, despite our picture of it as a lumbering, steaming hulk, is in fact an elegant epitome of change and summarizes the essence of irreversible change. The mid-nineteenth-century physicists were right to concentrate on it, for it focussed their attention on the crucial aspect of the irreversibility of the world.

That the heat engine captures the essence of the Clausius and Kelvin statements of the Second Law can be seen as follows. First, a heat engine needs a hot source and a cold sink, which is effectively Kelvin's statement. Moreover, in order to achieve the reverse flow of energy, from the cold sink to the hot source, the engine must be driven by an external motor, as in a refrigerator. The fact that work must be done in order to achieve this unnatural reverse flow is equivalent to Clausius's statement of the Second Law.

THE DISPERSAL OF ENERGY

Now the skin of the Second Law, which is a commentary on experience, will be peeled back to show its inner workings in terms of processes occurring on an atomic scale.

Everyone knows why things change: things tend to get worse. Scientists express this differently, less emotively, more precisely, and in a manner open to being rendered quantitative, by saying that *things tend to disperse*. All the Second Law is doing is to express outwardly the natural, purposeless tendency of things to disperse, where by things is meant (a) particles of matter, (b) energy, and (c) coherence. The last 'thing' is explained later. All three aspects of dispersal are really the same; it is only for convenience that they are divided into three classes.

The most primitive example of a natural tendency for things to disperse is the dispersal of the molecules of a gas. Everyone knows that molecules of gas accumulated in one region of a container will quickly spread throughout it. The reverse of this process, in which molecules regroup and collect in one small region, is unnatural, and has never been observed. It is particularly important to note that this chaotic dispersion of molecules is *purposeless*, nothing drives it in a special direction, and the destiny of the molecules is a consequence of their random jostling.

The second primitive process is the natural tendency of a hot object to cool to the temperature of its surroundings. This is a consequence of the dispersal of energy, as may be seen as follows.

A hot block of metal is a collection of randomly and vigorously moving vibrating atoms. The atoms at the edge of the block jostle the atoms in the surroundings and stimulate them into vibration at the expense of the vibration of the atoms in the block. As a result, the energy of the block jostles randomly and purposelessly out into the environment (fig. 6.2). The natural direction of change is the continuation of this dispersal of energy until it is distributed uniformly over the universe. Note once again that the dispersal is purposeless, and that it is a natural consequence of the universe being able to explore its possible states as time advances.

The picture of energy dispersing accounts for the Clausius version of the Second Law, because large spontaneous deviations from uniform dispersal are exceedingly rare. It is extremely unlikely that a significant fraction of the energy of the surroundings will accumulate spontaneously back into the initially hot object. The natural direction of change is for energy to flow as heat from a concentrated region and to become dispersed; the reverse of this flow is so improbable as to be regarded as unnatural. In other words, the natural direction of change is from hot to cold, in accord with Clausius's summary of experience.

We now turn to the third, slightly enigmatic form of dispersal, the dispersal of coherence. It can be introduced by considering another everyday irreversible process, that of a bouncing ball. Once again, we all know that as a ball bounces it gradually loses height. There have been no reliable reports of anyone seeing a ball resting quietly on the table suddenly beginning to bounce, and then bouncing higher and higher. This never-reported process is allowed by the First Law of thermodynamics (that energy is conserved) because the resting ball can acquire energy from the table, the latter becoming slightly

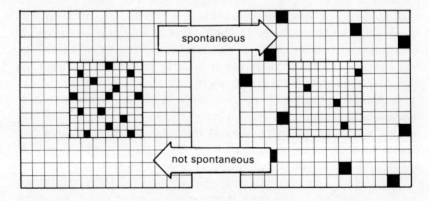

Fig. 6.2

cooler. However, the spontaneous bouncing is against our experience, and is against the Second Law.

The atomic reason for the fact of importance that a ball does not spontaneously begin to bounce is as follows. Not only must energy accumulate in the ball from the table, it must also accumulate so that all the atoms in the ball begin to move simultaneously in the same direction. When a ball is heated, the *random*, incoherent motion of the atoms that compose it is stimulated. When a ball is struck by a bat, energy is transferred to it in a *coherent* manner, in the sense that all the atoms of the ball pick up the uniform motion of the atoms in the moving bat, and begin to move in step.

By dispersal of coherence is meant the natural tendency for the orderly motion of atoms in an object that is moving to become corrupted into incoherent, random motion. Thus uniform motion has a tendency to decay into the random motion we call thermal motion. The sequence of events that occur when a bouncing ball strikes the table can be understood in terms of the loss of the coherence of the motion of its atoms, and the unnaturalness of the reverse process reflects the overwhelming improbability of coherent motion emerging spontaneously. Thus, the ball with initially all its atoms moving coherently strikes the table; the atoms of the table are jostled into random motion. On each successive bounce the dispersal is more complete, and finally the ball sits quietly on the table with all its atoms vibrating at random, with no net motion of the ball as a whole. The reverse sequence of events, in which all the trillions of atoms of the ball are initially vibrating at random, and then suddenly come into step with the result that it flies up into the air, is so unlikely that it can be dismissed as being impossible.

This picture of the dispersal of coherence accounts for Kelvin's version of the Second Law, for if it were not true, a device could be invented that went against his statement. For example, a ball resting on a table might suddenly leap higher at the expense of the energy in the table, which would slightly cool. But if the ball were to bounce higher it could be caught when it was high in the air, and then used as a weight to raise another weight. At the end of the process there would be a ball and a table in their original positions, but slightly cooler, and a weight would have been raised, which is contrary to Kelvin's experience.

In summary, the dispersal of energy and matter relates most closely to the Clausius statement of the Second Law, and the dispersal of coherence relates most closely to the Kelvin statement. It has already been mentioned that the three forms of dispersal are actually examples of a single process, and so at root the Clausius and the Kelvin statements can be expected to be equivalent. They can in fact be shown to be equivalent, but we will not pursue the proof here.

All the foregoing can be expressed in the form that:

> irreversible change arises from purposeless drifting into the available states.

In the case of cooling, the drift is the dispersal of energy; in the case of matter, the drift is the dispersal of particles; in the case of the loss of net motion, the decay is in terms of coherence. So far, of course, there is no memory and therefore nothing that can be called an experience. In order to notice the passage of time we need to form memories, and memories (in living systems) are changes of chemical composition in the cells of the brain. It is the purpose of this chapter to show that such complex change occurs by a straightforward elaboration of the purposeless dispersal just described, and in order to set the trap to capture this form of change, it is necessary to expand and elaborate what has been said so far.

ENTROPY

The thermodynamic property called *entropy*, which is denoted by the letter S, is introduced so that the intuitive notions relating to dispersal can be made quantitative and extended to include chemical processes.

The thought that motivates the introduction of entropy can be expressed as follows. The First Law of Thermodynamics leads to the introduction of a property, *energy*, that can be used as a convenient label to distinguish states of the universe that can be reached from a given initial state from those that cannot: only states with the same total energy are *attainable* states because energy can be neither created nor destroyed. Thus, if two states of the universe have exactly the same energy, then it is possible in principle for the universe to make a transition between them. The energy of the universe is a label that is used to assess whether a certain change is feasible. Similarly, the entropy is a numerical property introduced to distinguish states of the universe that can be attained *spontaneously* from other feasible states that cannot. Thus the value of the change

of entropy for a process determines whether the proposed final state
is spontaneously accessible from the proposed initial state.

There is nothing mysterious about entropy, for it is just a
convenient ordering number. Entropy's only odd feature is that it is
defined so that its value *increases* in a natural change. Hence, if two
states of the universe are considered, and those states have the same
total energy (so that transitions between them are feasible), and one
state has a higher entropy than the other, then the direction of
spontaneous change of the universe is from the state with low
entropy to the state of high entropy. A summary of the role of
entropy is therefore the *entropy principle*, that:

>natural changes occur in the direction of increasing total entropy.

Since we have already seen at some length that natural change
occurs in the direction of dispersal, it follows that entropy must be a
measure of dispersal. This is the link forged in the next section.

CHANGES OF ENTROPY

The following (slightly imprecise) definition of an entropy change is
compatible with everything expressed so far, but goes beyond by
making it precise and quantitative:

$$\text{Change of entropy} = \frac{\text{Heat transferred}}{\text{Temperature}}$$

Heat is an energy transfer in which atoms are stirred into chaotic
thermal motion; if a lot of energy is transferred as heat a lot of chaos
is generated. This is consistent with *heat transferred* appearing in
the numerator of this expression. Energy can also be transferred as
work, but this has no effect on the entropy because it creates no
turmoil: hence the entropy change is defined in terms of the energy
transferred as heat, not as work. The presence of the temperature in
the denominator is consistent with there being a greater increase of
chaotic turmoil when the energy is transferred at low temperatures
(when there is little thermal motion present) than at high (when
there is already much thermal motion): a sneeze in a library creates
much more additional turmoil than a sneeze in a busy street.

That the definition of entropy change is consistent with the entropy principle can be checked by considering the Clausius and Kelvin statements.

Figure 6.3 shows the transfers of energy that would be natural if the Clausius statement were false. For a transfer of a certain amount of energy as heat, the entropy of the cold source falls because energy is transferred from it as heat. The same energy arrives as heat in the hot sink, and increases the latter's entropy, but by a smaller amount on account of its higher temperature. The overall effect is therefore a decrease in the entropy of the universe, which is against nature, as the Clausius statement asserts.

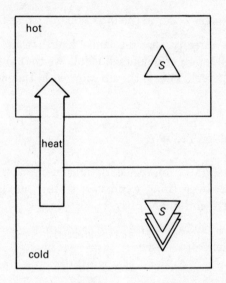

Fig. 6.3

Figure 6.4 shows a similar argument for the Kelvin statement. The entropy of the hot source falls because energy is removed from it as heat. That energy is converted, using the engine, *entirely* into work. Hence the overall effect is to withdraw energy as heat (resulting in an entropy decrease) and to transfer it to the surroundings as work (resulting in no change of entropy). Hence, overall the entropy falls. This process is therefore unnatural, which is Kelvin's assertion.

Figure 6.5 shows that the entropy principle and definition permit the extraction of work from a heat engine. In this case the entropy falls on account of the withdrawal of energy as heat from the hot

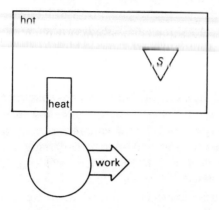

Fig. 6.4

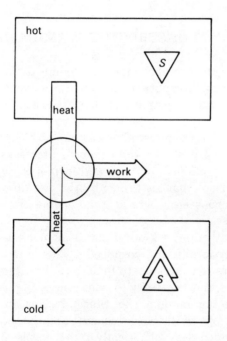

Fig. 6.5

source. If all that energy were to find its way to the cold sink, there
would be an increase in entropy because the deposit is being made at
a lower temperature. In fact, only some of the energy need be

deposited as heat for there to be an overall increase in entropy, and the balance of energy may be withdrawn as work. The minimum amount of heat that must be deposited into the cold sink can be calculated very easily, and has important consequences for the efficiencies of all kinds of heat engine.

The introduction of entropy gives a convenient way of expressing the underlying tendency of the universe towards chaos and the tendency of energy, matter and coherence to disperse. Instead of drawing diagrams to represent energy dispersing, a somewhat vague concept, we can deal with the numerical values of the entropy and draw diagrams showing how the entropy increases, which is a very precise, quantitative concept. From now on we use the increase of entropy as signifying the direction of change, but its underlying basis, purposeless dispersal, should not be forgotten.

THE EMERGENCE OF STRUCTURES AMID CHAOS

How can it be, if all natural change corresponds to the collapse of our universe into cosmic corruption, that any improvement can emerge? What accounts for the emergence of an occasional cathedral, a person or an opinion?

The answer is that the physical processes on which the construction of a cathedral depends, like the chemical processes on which the formation of perceptions, opinions, and memories depend, are *local* abatements of chaos that are driven by the generation of more than compensating chaos elsewhere so that *overall* the entropy of the universe increases.

Building a cathedral, a kind of social memory, is going against nature, but only locally, not cosmically. This is represented in fig. 6.6, which shows that in order to build a rudimentary cathedral, let alone a Chartres, it is necessary to raise randomly placed blocks into the appropriate non-random pile. Since there is an accompanying reduction of entropy of the blocks constituting the cathedral (because they are arranged less chaotically), cathedrals do not emerge spontaneously. However, a compensating amount of entropy can be generated by coupling an engine (including well-fed men) to the blocks. There is now an increase in universal entropy on account of the flow of energy from hot to cold, and this increase may be enough to overcome the local loss of entropy as the blocks are raised into position. The entropy of the universe has increased even though

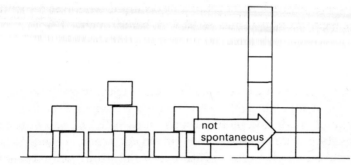

Fig. 6.6

locally a cathedral has been built and corresponds *locally* to a diminution of chaos. Exactly the same kind of argument relates to the chemical formation of a memory, as we see shortly.

A second example of apparently, but in fact only locally, going against nature is refrigeration. In refrigeration we seek to contrive to go against Clausius and to transfer energy as heat from cold to hot. This is represented in fig. 6.7, where the process on the right shows

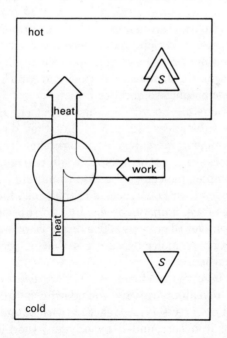

Fig. 6.7

the flow of energy which, if that alone occurred, would correspond to a decrease of entropy since energy is being removed from a cold source and being deposited in a hot sink. But if energy is added to the stream, then more energy can be deposited as heat in the hot sink than was removed from the cold source (the interior of the refrigerator) and hence an overall increase in entropy can be achieved. This is the strategy in a refrigerator: the flow of energy is augmented by coupling an electric motor, which adds to the stream of energy and therefore, if it is powerful enough and adds enough energy to the flow, results in an increase of entropy overall. However, in order for the refrigerator to run against nature in the kitchen, it is necessary that somewhere else at least a compensating amount of chaos must be generated to produce the electricity that runs the motor that adds to the stream of energy that results in an increase in entropy overall. That is why there needs to be a connection of the refrigerator to a waterfall, a furnace, or a reactor elsewhere: the national grid connects users to centres of generation of chaos.

CHEMICAL REACTIONS

The preceding remarks are simplified descriptions of physical processes, processes in which matter does not change from one substance into another. In order to penetrate the brain it is necessary to switch attention to chemical reactions, in which matter does change from one substance to another.

Chemical reactions are heavily disguised heat engines. Reactions proceed and new substances are formed, but they do so *only if the entropy of the universe increases*. Although chemical reactions do not rearrange blocks into cathedrals, they do rearrange atoms into molecules, for complex molecules are akin to minute cathedrals. The principle that drives their construction is the same, for there is only one principle of change, namely descent into corruption (or, in other words, increased universal entropy). The dispersal of energy accounts for the construction of molecules no less than it accounts for the construction of cathedrals.

The processes involved in chemical reactions can be expressed in an abstract but revealing way by considering as an example the rusting of iron (more precisely, its reaction with oxygen). Figure 6.8 shows a block of iron surrounded by oxygen; the block might be shaped into something resembling a car, and the oxygen may be that

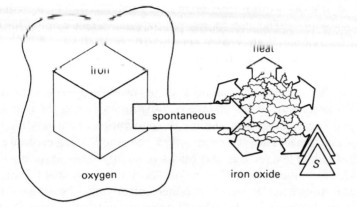

Fig. 6.8

present in the atmosphere. When oxygen combines with iron a great deal of energy is released as heat, and it dissipates into the surroundings, increasing their entropy. The reaction of iron is therefore a chemical analogue of the cooling of hot metal. In a sense, rust is chemically 'cooler' than iron.

The entropy of the universe increases as rusting proceeds, hence the reaction is spontaneous, and all iron has a natural tendency to turn to rust. The rusting can be represented as shown in fig. 6.9,

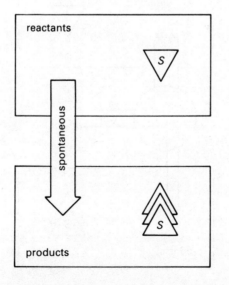

Fig. 6.9

which stylizes a chemical reaction in a way that shows its resemblance to a heat engine. The 'hot source' is now an abstract symbol for the 〜〜〜〜〜 and the 'cold sink' 〜〜〜〜〜 〜〜〜〜 〜〜〜〜〜〜 As in any process, the reaction is spontaneous if the flow from reactants to products is accompanied by an increase in universal entropy, an increase that arises partly on account of the energy released as heat but also as a result of the change of substances during the reaction. Moreover, as in a real heat engine, the generation of entropy may be more than is necessary to ensure that the reaction is spontaneous, and instead of extracting the released energy as heat it may be possible to channel some off as work. As shown schematically in fig. 6.10, the energy not needed to produce entropy may be drawn off as muscular effort or used to build the molecules involved in the physiology of perception, reflection, and memory. The abstract structure shown in fig. 6.10 is mapped into reality by taking the 'hot source' to be our food, the engine to be our body and brain, and the 'cold sink' to be our excrement. The cells of our body are the engines that harness collapse and convert it to an action, an opinion, or a memory.

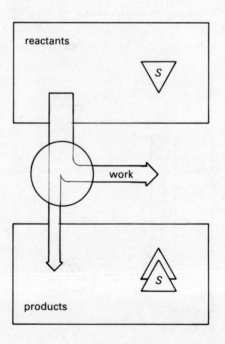

Fig. 6.10

THE STREAM OF CORRUPTION

What is the origin of the food? It could be milk produced by a cow, so it comes further up the stream of corruption. This is represented in fig. 6.11, each stage of which shows a pair of coupled chemical 'heat engines'. Energy in excess of that needed to produce enough entropy to ensure spontaneity in one reaction, its so-called *free energy*, is being used, just as burning fuel is used to refrigerate elsewhere, to produce structure in another reaction, and in the subsequent box that structure is used as fuel to drive the next reaction, and so on: grass is used to produce milk, then milk is used to produce thoughts.

The chain can be traced back to the very low entropy radiation of the sun, and its spontaneity can in turn be traced back to the extremely low entropy energy of the big bang. Thus, when we are perceiving, effecting purposeless action, and memorizing, we are actually participating in a highly geared cosmic unwinding. The purposeless dispersal of energy leads to apparently purposeful action and intricately modified brains, such is the complexity of the gearing through which the universal collapse occurs.

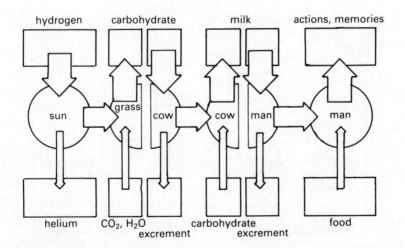

Fig. 6.11

PROCESSES OF EXPERIENCE

One way of tapping the energy in order actually to produce creative acts, perceptions, muscular effort, and memory is through electrochemistry, and evolutionary pressures have led to the deployment of eletromechanical processes to power the activities of living things.

Electrochemistry, and its relation to entropy, can be discussed in the following terms. Figure 6.12(a) shows a process in which zinc is dissolved in copper sulphate solution, with copper depositing on the surface of the zinc. The fact that the reaction occurs spontaneously indicates that it is accompanied by an increase of entropy, but we will not pursue the detailed reasons for the increase. The reaction can be expressed in terms of the flow of electrons at random from the zinc to the copper ions in solution, with the result that copper atoms are formed, and zinc atoms are converted to ions and dissolve. Notice that there is a *random* electron flow and that the energy of the reaction is released as heat.

Now suppose that the zinc and copper are separated as in fig. 6.12(b). The same overall natural process occurs, but the electrons released by the zinc must travel through the external circuit to the copper. This orderly flow of electrons through the circuit is an

(a)   (b)

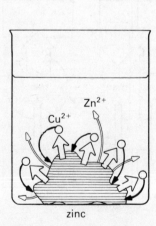
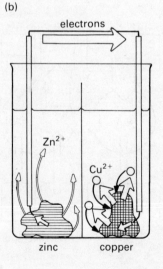

Fig. 6.12

electric current, and it may be used to drive an electric motor. That is, instead of releasing its energy as heat, the reaction can be used to do work. This is the origin of the action in all electrochemical devices: the device functions because it harnesses the entropy increase (the dispersal of energy) to drive electrons through a circuit, and the local abatement of chaos is the work that can be extracted from the electron flow.

The particular electrical device to have in mind is the human body, and that particular component of the body, the brain. The electrochemical processes described above also occur in our bodies, the only difference being that we live off food and air, not copper and zinc. However, as for the reaction between copper and zinc, an electron flow arises from the process of energy dispersal, with the universe drifting into greater chaos. In fact, all the processes in our cells, including our brain cells, can be regarded as a collection of little electrochemical cells (fig. 6.13), with each one building a little molecular structure as it runs irreversibly in the direction that adds to the universal chaos. Those structures are the modifications of our brain that leave us with our memories, our experience of time.

We feed (ultimately off the sun, and further back in time), and, as we feed, processes occur that build other molecules, and under the influence of internal and external influences, the processes build molecules that store memories. The imprints of change left in our brains are the footsteps of the chemical reactions that have occurred

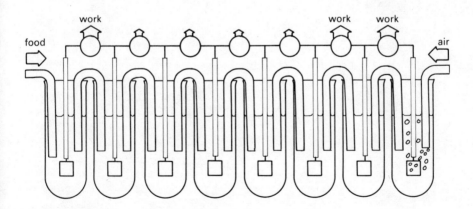

Fig. 6.13

and that are left imprinted as a memory to modify our later actions and opinions.

We have looked through the window on to the world provided by the Second Law, and have seen the naked purposelessness of nature. The deep structure of change is decay, the spring of change in all its forms is the corruption of the quality of energy as it spreads chaotically, irreversibly, and purposelessly in time. All change, and time's arrow, point in the direction of corruption. The experience of time is the gearing of the electrochemical processes in our brains to this purposeless drift into chaos as we sink into equilibrium and the grave.

FURTHER READING

Atkins, P. W., *The Second Law* (Scientific American Library, W. H. Freeman, New York, 1984).

Fenn, J. B., *Engines, Energy and Entropy* (W. H. Freeman, New York, 1982).

Peacocke, A. R., *An Introduction to the Physical Chemistry of Biological Organization* (Oxford University Press, Oxford, 1983).

Prigogine, I. and Stengers, I., *Order Out of Chaos* (Heinemann, London, 1984).

Stryer, L., *Biochemistry* (W. H. Freeman, San Francisco, 1981).

Zemansky, M. W. and Dittman, R. H., *Heat and Thermodynamics* (McGraw-Hill, New York, 1981).

7
Time Asymmetry and Quantum Mechanics

PAUL DAVIES

It has fallen upon me to give what I think is the most difficult of the talks of this series because, as I am sure you are only too willing to agree, quantum mechanics is not the easiest of subjects even for professional physicists. Its deeper aspects concerning measurement are by no means thoroughly understood either by physicists or by philosophers, and there is no general consensus among these groups of people on which particular interpretation to adopt. However, I think one point is agreed among physicists and philosophers and that is that the act of measurement or observation in quantum mechanics is necessarily associated with irreversibility or some time-asymmetric change. It is that irreversibility which I want to focus on, both in order to explain what it is and also to try to bring out the connection, which I believe exists, between time asymmetry in the act of quantum measurement and other aspects of time asymmetry in physics.

Bohr said that no elementary quantum phenomenon *is* a phenomenon until it has been brought to a close by an irreversible act of amplification. The key feature here is that if we are observing or measuring a quantum system, which is usually something very small, for example the position of an electron, then at the end of the day (to use a rather overworked expression) we will normally want to discuss something big, macroscopic and uncontentious, like a click on a Geiger counter. The act of going from the microsystem, which, as we shall see shortly, is a fuzzy, rather indeterministic thing, to this big macrosystem, which we all feel is in some sense a well-defined state with a concrete existence, is clearly an act of amplification. But the crucial point is that it is also irreversible. Bohr regards this irreversible act of amplification, the stage at which one can say that a phenomenon has come into being, as in some sense generating reality. I don't want to get too far into the issue of reality in quantum physics because it is a difficult one with many ramifications, but at least in Bohr's mind the point at which one can say that concrete reality has been achieved is the point at which one can say there has been some irreversible change. This places time asymmetry in a very fundamental position with respect to our description of the world. It seems to imply that it is tied up in an intimate way with our very notion of what it is for something to be real.

In daily life, as we all know, there is no doubt that the world is time-asymmetric. We see many, many phenomena about us which are, crudely speaking, directed one way in time. The physicist likes to describe time asymmetry or irreversibility in terms of the concept of *entropy*, which is a much-used word that can roughly be associated with the notion of disorder. The so-called Second Law of Thermodynamics embodies a fact which I think we are all quite prepared to believe, namely that every day the universe gets a little bit more disorderly. When Humpty Dumpty fell off the wall he succeeded in disordering himself really rather badly and, even with the attention of all the King's horses and men, could not be reconstructed. Events like this, time-asymmetric events where entropy increases, are very familiar to us in daily life and in the laboratory. When it comes to quantum physics, however, we enter a world where things are not quite so clear cut. It is going to be something of a challenge to me to attempt to present some of the essential ideas of quantum physics in just a few pages. Usually, even after an entire lecture course, students are completely baffled! In fact Bohr said that anyone who is *not* shocked by quantum mechanics hadn't understood it. At any rate, I want to give a thumbnail sketch of at least the key features and concepts.

It's helpful to go back almost to the time when quantum physics came into being, which was with wave-particle duality. Most people are at least vaguely aware that something funny happens when we start probing around at the atomic level. Systems such as photons and electrons have both particle-like and wave-like aspects; and there is something a bit paradoxical about this. For example, I think most people are prepared to think of an atom, or better still an electron, as a particle. It's most compelling to think of it as just a scaled-down version of something like a billiard ball and attribute to it all the sorts of things we would like to attribute to macroscopic objects, such as having a location and a motion and so on. Conversely, one is used to thinking of light as an electromagnetic wave and having all the attributes that one associates with a macroscopic wave, such as being spread out in space and carrying energy and so forth.

Unfortunately, when we go down to the microscopic level there seems to be present an element of both wave- and particle-like behaviour. So, for example, electrons can behave under some circumstances as waves; and equally, light can come in lumps or units called photons. The problem is: how can something be both a wave and a particle? As a result people talk about *duality* as a convenient means of discussing this phenomenon. They simply say that there are some experiments in which the wave-like character is manifested and some in which the particle-like aspect is manifested. However, students always insist on asking what it *really* is. It is at this point that opinions start to differ. I tend to follow most of the way along what has become known as the official view, the party line, or the *Copenhagen interpretation*, which is primarily due to Niels Bohr. On this view you cannot talk about what is really there, or rather you can talk about it but you can't make much sense of it, until you specify the context of the experimental arrangement. That is to say, if you specify what you are going to measure, for example the position of an electron, then you are perfectly entitled to carry out the measurement, find the result and say that there is an electron with position such-and-such. Or equally you can carry out a measurement of the momentum of the electron and find a value for the momentum. Then you have an electron with a momentum. But since you can't perform both experiments simultaneously, it is held to be simply meaningless to talk about the electron really having a position or really having a momentum independently of the experimental context.

That's my interpretation of the official view; and we have the famous Heisenberg *uncertainty principle* which asserts this inherent incompatibility between the measurement of the position x and the measurement of the momentum p. You can know one or the other individually, to any desired degree of precision, but not both, and there will be an irreducible uncertainty in the results of the measurements given by the formula

$$\Delta x . \Delta p. \gtrsim h/2\pi$$

Δx is the uncertainty in the position while Δp is the uncertainty in the momentum, and h is Planck's constant, which gives the scale of the phenomena that I am talking about. So there is this inherent uncertainty.

Einstein never liked these ideas because he wanted to believe in some sort of naive reality, to believe, so to speak, that the universe is out there ticking over independently of our observation. Bohr's position, by contrast, is that the observer and the observed are folded together in a very intimate way and that they can't be separated out clearly into one and the other. The nature of what is out there, according to Bohr, is inextricably mixed up with what the observer chooses to observe or how the observer chooses to discuss the world, that is, the experimental context. This is very baffling and very mystical, although a lot of people, of course, delight in that because they like to find mystical things in physics and they often go off to India to contemplate these matters. The fact remains, however, that the standard textbook position on quantum mechanics is that one simply cannot ascribe a full set of attributes to a particle such as an electron in the same way that you would to a particle such as a billiard ball. In other words, electrons differ in a fundamental respect from things like billiard balls not only in scale but also in the way in which we can ascribe to them classical attributes. Consequently, there is an element of uncertainty and indeterminism, because of course if we don't know where something is, or if it's a bit fuzzy, then we can't be sure of how it's going to behave or develop, and so there is some element of indeterminism about the future. All we can do is to give the betting odds. We can talk in statistical terms about the probabilities of certain outcomes, of certain types of atomic activity, but we usually can't be certain. Hence the indeterminism.

Now there are two quite distinct types of uncertainty in nature. One of them is like the uncertainty of the stock market, or the uncertainty of the weather, where we believe that all the little fluctuations have their causes in something. It's simply that they are such complex systems that we can't follow all the little twists and turns in the detail necessary to predict their behaviour. However, the Copenhagen interpretation of quantum theory states that the indeterminism or the uncertainty in a quantum system is intrinsic to it and not simply the result of our having an incomplete knowledge of the system. It really is fundamentally, intrinsically uncertain or indeterministic.

When we come back to the problem of whether something is a wave or a particle, then the word *complementarity* comes in, because Bohr's position is that these are complementary descriptions of reality or of the system, and that they don't contradict each other. You can measure either the position or the momentum but not both simultaneously. So they give complementary descriptions of the world and not contradictory descriptions. You must specify what it is you want to talk about, what you are going to measure; and you can't talk about what reality is independently of such a specification.

The fact that a particle doesn't have a well-defined position and momentum simultaneously means that you cannot really talk about where it is going. In classical physics we like to think of a particle following a well-defined trajectory. For example this might be a bullet travelling to some target or a planet travelling in an orbit. In cases such as these, we assume, when a particle arrives at its destination, that it can in principle be connected back continuously along some well-defined path through space to its point of departure. But in quantum physics that cannot be so because to specify such a trajectory we would have to know both the position and the velocity of the particle at each point and we cannot do that. We can either know where it is or how it is moving but we cannot know both. John Wheeler has a very nice analogy here: he says that the motion of a particle is like that of a great smokey dragon with a sharp tail and sharp teeth. It can bite at the point of departure and at the point of arrival but in between all is fuzzy and we do not know what is going on.

This idea of not knowing where a particle is going is exemplified in a very famous experiment which gets to the very heart of the problem in quantum physics. It is an experiment which was devised by Thomas Young, Egyptologist and physicist, to demonstrate the wave nature of light (see fig. 7.1.) It consists of three components: a screen with two apertures, an image screen and a point source of light which illuminates the apertures. The image of the two apertures will appear on the image screen. At first sight you might think that if there are two holes in the screen then the image will consist of two patches of light beyond the holes. But in fact what we see is a series of light and dark bands, called *interference fringes*. This comes about because the light waves from each aperture interfere with each other

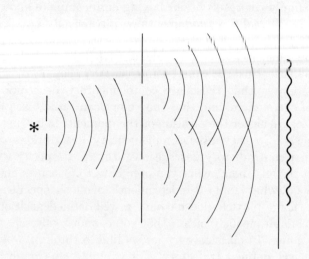

Fig. 7.1 *Young's experiment*

after they have passed through. The waves begin to overlap and where they arrive in step they reinforce each other and where they arrive out of step they cancel each other. Accordingly, we see bright bands where they reinforce and dark bands where they cancel.

The problem comes when we try to think of light as a stream of photons, particles that is, and in particular when we envisage turning down the strength of the source of light so that only one photon goes through the system at a time. Surely, we think, each photon must either go through one slit or through the other but not through both. Since the photon certainly arrives at the screen, we could put a photographic plate in the position of the image screen so that each photon would make a little dot on the plate. Then another photon would come through and we would see another dot, and so on. After a while, a pattern would begin to build up in this speckled way and we would obtain our fringes. Of course each individual photon can only make a spot; it cannot make a pattern. Nevertheless, each individual photon contributes to the pattern, so there seems to be a coherence, or perhaps a better word is a conspiracy, between the activities of all these different photons to go to the light fringe areas and not the dark fringe areas. Now the mystery is this: the interference pattern can clearly only occur if there are two apertures present, because the whole idea is that waves from one aperture get in the way of waves from the other. If we only had one aperture we couldn't have an interference pattern. Yet if each photon only goes through either one or the other, how is it that the presence of both slits can contribute to the behaviour of each individual photon? That is the mystery.

This, in fact, was one of the experiments that Einstein chose to attack, because he didn't like quantum mechanics. He asked what would happen if we set up an experiment to see which slit the photon goes through. We can do this by freeing the screen containing the slits to see how the photon gets deflected; if it goes through the upper slit it will be knocked upwards, if it goes through the lower slit it will be knocked downwards. There are also other ways of determining which slit the photon passes through. What Bohr then showed was that when we do the experiment we wash out the interference pattern. In other words, Nature outmanoeuvres us. We are dealing not with contradictory but with complementary descriptions. We have here two complementary experiments. In one of them we don't look to see which route the particle took; we can think of the particle as probing both of the slits. In the other experiment we do look at the route but we don't get interference patterns. These are complementary descriptions of the world. They don't contradict each other, but we do have to decide on our experimental context before we can say whether the photons will have gone through a particular slit or not.

One way of looking at this is in terms of two completely different worlds, two potential worlds or maybe even two actual worlds. In one world the particle goes through the upper slit and in the other it goes through the lower slit. As long as we don't look to see which route the particle took — as long as we have this smokey dragon — we can say that in a sense both of these worlds co-exist. Furthermore they overlap each other and cause interference effects. But we are now getting very close to the question of irreversibility, because there are many examples of quantum systems where we overlap two essentially different worlds, or two waves, each representing a different world. When we choose to look at either one or the other we always get a definite result. This transition from the fuzzy, smokey dragon to the sharp, concrete world of daily experience of macroscopic things is sometimes referred to as the problem of the collapse of the wave function.

There are many reasons why one can think of it as a collapse. Let us, for example, suppose that the wave in some way encodes the position of the particle. If the wave is spread out over a certain region of space then we are not sure of the precise position of the particle in this region. However, we can choose to measure the particle's position. Then the wave will collapse down to some very sharp-peaked thing in the region where we made the measurement. That enables us to say that the particle is at such-and-such a position. Now this collapse of the wave function is an irreversible collapse; it is something that we cannot undo. In fact one can be quite explicit about this and formulate mathematically an expression for the entropy of the quantum system. This entropy definitely increases after the measurement has taken place, so it is an irreversible change. The question is: can one in some way relate this jump in the entropy of the microsystem or quantum system to the macroscopic expression for the entropy of the world as a whole?

Let me give you an example of the problems that arise with this collapse of the wave function. Suppose we have a particle confined to a box — this is a nice simple thing that physicists like to talk about. Classically, according to Newtonian mechanics, the particle is simply rattling around somewhere in the box. If we choose not to look and see where it is then we can regard every position inside the box as equally probable. In terms of quantum mechanics we have a rather similar sort of description but this time we think of the particle as being represented by a wave. The wave fills the box uniformly and again we would say that there is equal probability of finding the electron, or whatever particle it is, at any point inside the box.

Now suppose we try something a bit more elaborate. First we put the particle in the box, but then we come along with a membrane and divide the box in two. The particle is in there somewhere, either on the left or on the right; it cannot be in both chambers. But we know that the act of inserting the membrane chops the wave in two, so there are still waves on the left and on the right. Nevertheless, the particle must be either on the left or on the right. How are we to decide what is going on? If we have an observer who chooses to open the box then we can get some information about what is happening. Suppose the observer opens the box and finds that the particle is in fact in the left-hand chamber. Then abruptly the wave in the right-hand chamber disappears. The wave function collapses into one half of the box. What bothers people is that the collapse seems to be instantaneous. Those of you who follow the theory of relativity will know that the idea of simultaneity is a vexed issue and we would not like to suppose that any physical effect propagates instantaneously across the universe. So we have the problem of reconciling the collapse of the wave function with the theory of relativity. It turns out, however, that this problem is not insuperable. There is actually no conflict between the two ideas. But I cannot pursue these matters here. For I want to return to the idea of irreversibility that we have been considering so far.

The description of this superposition of worlds, or superposition of waves, is quite a straightforward one in quantum mechanics. We can, for example, imagine having two waves, each one representing a possible state of a photon. Let us think of them as a red wave and a blue wave. Continuing the optical analogy, it is quite usual to obtain a superposition of these two wave forms by simply bringing them together. Given any two wave forms, we can obtain more complicated wave forms consisting of these two convoluted together. Mathematically, we can describe this by saying that the total wave is the sum of the other two waves with a certain amplitude for each (see fig. 7.2).

In quantum mechanics these two waves could represent two possible different outcomes resulting from a measurement. For example, one wave could be the photon going through one slit while the other wave might be the photon going through the other slit. If we choose to make the measurement of which slit the photon went through, then of course we get a result. The fact that we obtain a particular result means that the wave corresponding to the photon going through the other slit abruptly disappears. This again is the collapse of the wave function, with the wave which consists of these two pieces together suddenly jumping down either to the one or to the other. The inherent unpredictability of the quantum system means that we cannot say in advance which wave will be obtained. But we shall certainly get either one or the other and not both. So the collapse of the wave function, which is this key irreversible step, is associated with going from an overlapping superposition of worlds either to one concrete world or to the other. It is a little bit as though one's got a ghostly coexistence of two realities (I shouldn't really use the word 'reality' but I can get away with it just this once!) and that the act of measurement has the effect of projecting out either one or the other. Furthermore, this projection is an irreversible thing associated with a real increase of entropy. Returning to our red/blue analogy, we can think of the two worlds — the red world, represented by the red wave and the blue world, represented by the blue wave —

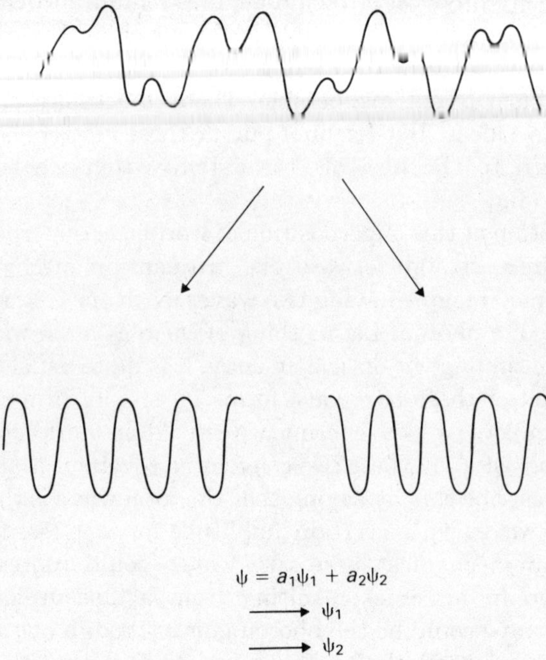

$$\psi = a_1\psi_1 + a_2\psi_2$$
$$\longrightarrow \psi_1$$
$$\longrightarrow \psi_2$$

Fig. 7.2

as coexisting and overlapping. I must emphasize the overlapping. If they just coexisted but didn't get in each other's way, then we could think of them as two disconnected realities not interfering with each other. However, they do interfere and overlap and so we cannot ignore them.

There are many different interpretations of what is going on here in the act of measurement and the associated projection, and I shall be coming to them in a moment. If we leave a quantum system to itself, the wave function — the thing that represents the wave amplitude of the system, the photon say — evolves in time in what the physicists call a *unitary* fashion, which, crudely speaking, means in a reversible fashion. By leaving it to itself I mean choosing not to make a measurement or not to look. When we make a measurement, however, the wave function changes in a completely different way, abruptly and irreversibly. So the act of making a measurement on the system is to cause it to evolve in time in a characteristically different way from when we choose not to make a measurement.

Some of you may have already leapt to a mystical conclusion and will think that this is surely mind over matter, because if we can show in a laboratory that when we observe something we make it jump, isn't that mind acting on matter? Well, you have got to be very careful, because even in classical physics carrying out an observation inevitably involves some sort of coupling between the observed system and the observer, and every sort of coupling will involve a disturbance and so it is not surprising that when you observe something you disturb it. If you want to measure the temperature of a cup of tea you've got to put a thermometer in it and of course that will change its temperature a little bit. The magnitude of the change in temperature is not the point. The point is rather that this act of coupling the observed system to the observer is something which in classical physics could be regarded as merely an incidental feature of the observation with no fundamental significance. It could in principle always be reduced to some very small and unimportant disturbance. In quantum physics, on the other hand, it is absolutely fundamental and cannot even in principle be reduced to something small enough to be ignored.

Some people, certainly Hermann Bondi but I think it goes back to Eddington, have suggested that this actual collapse, this irreversible change of the wave function, is responsible for our impression of the flow of time. I don't actually support this. In fact I would be prepared to argue that our sense of the flow of time is purely an illusion, like our impression, when giddy, that the world is whirling about us, and I think they have a similar status. However, that's taking us off the subject of the lecture. You might suppose that all of this argie-bargie with collapsing worlds and collapsing wave functions really does not matter very much seeing that it all takes place down there among the atoms; since they are so small and inaccessible it can't have any importance for the everyday world. This is not so, however. For there are many ways in which we can amplify this ghostliness, or this superposition of worlds, from the microscopic, the microcosm or microcosmos, up to the world of everyday proportions.

Here we arrive at the famous Schrödinger cat paradox, which is simply one means of attempting to amplify this superposition of worlds to the macroscopic condition. I call it the paradox of observation. Let me first of all explain what fig. 7.3 represents. I have attempted to show that when we are talking about, for example, electrons or photons, there is an inherent uncertainty or unpredictability in their behaviour, and that the situation may be such that we can envisage two overlapping superimposed worlds that interfere with each other, one representing the electron or photon doing one thing and the other representing the electron or photon doing something else. If, through some device such as a Geiger counter, we were to couple the microscopic system up to some macroscopic piece of apparatus, like a hammer, then it would be possible to take this superposition of two microscopic worlds and turn it into a super-position of two completely different macroscopic worlds.

The way in which Schrödinger proposed that this should be done was to incarcerate an animal, a cat in fact, into a box with this device which amplifies from the microscopic to the macroscopic. The hammer in the box can smash a flask of cyanide or something of that sort. The experiment consists of the following. The microscopic system, which might be the nucleus of an atom, is placed in a superposition of states such that one state will trigger the hammer and the other state won't. You could, for example, arrange that if the atom decayed after a certain time, say one minute, then the hammer will fall, and if it didn't decay, then it would not. Whatever the details, the point is that we have two completely different worlds because in one the cat

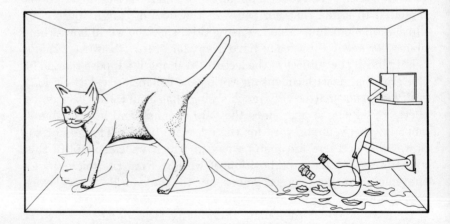

Fig. 7.3 *Paradox of observation*

remains alive and in the other it is dead. So while this formula (fig. 7.2) seems all right for atoms we begin to wonder if we have not taken leave of our senses when it comes to pussy being in a superposition of states, $a_1 \psi_1 + a_2 \psi_2$, such that ψ_1 corresponds to being alive and ψ_2 corresponds to being dead. Isn't this complete nonsense? For what, after all, are we to make of a live-dead cat? At the very least, we think, the cat itself must know whether it is alive or dead (or at any rate must know if it's alive).

(I once had occasion to make a television programme about this whole subject and took the opportunity symbolically to re-enact the Schrödinger cat paradox with a real live cat which one of the cameramen had produced. It was an extremely docile creature at the start of the experiment, but after we'd had about fifteen retakes stuffing the cat into the box and forcing the lid down, it had become totally and utterly neurotic! So when it came to the final take it insisted on miaowing all the way through the process and thus collapsing the wave function all the time. It was very difficult, at the end, for the viewers to tell whether it was alive or dead, because when we lifted the lid it shot out so fast that I don't think any of the viewers could spot it. The moral of that is: never do television programmes with children or animals!)

The resolutions of the Schrödinger cat paradox, and these other peculiar features I have been talking about, are many and much discussed. I mentioned at the start that there is no consensus among physicists, and certainly not among philosophers, as to how we should extricate ourselves from these difficulties.

I suppose I should stress that there is no problem as far as the actual implementation of quantum mechanics is concerned. By this I mean that any physicist is perfectly happy to turn the handle and use quantum mechanics as a procedure or an algorithm for computing the results of actual experiments. There are no serious doubts cast upon whether quantum mechanics is correct as far as the formal structure of the theory is concerned. The problem is rather that, when taken at face value, the theory doesn't seem to make much sense. Consequently, it is always necessary to graft some sort of epistemological assumption onto the bare formalism in order to extract any sense from it. These assumptions vary according to taste. For example, it has been suggested by a very small minority of physicists, most notably Eugene Wigner, that we will not escape from the paradoxes of quantum physics unless we are prepared to entertain the idea that mind, in some sense, has a fundamental role to play here. According to these physicists, if the act of making a measurement collapses the wave function then we cannot avoid talking about consciousness at some stage in the argument. Now it is certainly true that when the physicist writes down this wave function (I've been a little vague about what it means and how one processes it), he has in mind that it encodes what we know about the system. When an observation is made the wave function changes abruptly because our state of knowledge is changed; and if we are talking about knowledge then presumably we have to talk about someone there who knows. So it does seem, on following this line of argument, that one has to introduce mind into the proceedings at some stage. This, I imagine, would be anathema to most physicists. But I mention it because it is certainly an interpretation that has been proposed. We would then identify the time asymmetry with the entry into the consciousness of the observer of the details of where the electron is or, perhaps, how quickly it is moving. With this view we can see in a very clear way how the nature of reality is connected with the time asymmetry in the quantum measurement. For it is the very entering into the consciousness of the individual which is associated with that collapse of the wave function or that irreversible step.

Another approach is the *many worlds* or *many universes* interpretation. This seeks to accept in a sanguine way the existence of all these multiple realities. I have been talking about two possible worlds. For example we might have a photon going through one of two slits or we might have a live or a dead cat. These are normally regarded as alternative potential worlds and when you actually make an observation you collapse this ghostly superposition onto either one world or the other. That is the usual position, but in the many universes intepretation these are not potential worlds but actual worlds which actually exist. In fact they coexist. They are in parallel and so people sometimes talk about the *parallel universes* theory. On this view, every time we set up an experiment where we present an electron with a choice of alternative outcomes, we are inevitably multiplying the number of possible existing worlds. This is sometimes called the *branching universes* theory, and we can depict this simply by a tree in which we imagine that the world at this instant is splitting and re-splitting into countless copies of itself (see fig. 7.4). People argue about how many there are. There are certainly an infinite number of copies, because in most experiments the particles are presented with an infinity of choices and not just two. According

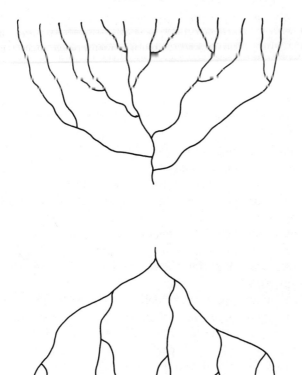

Fig. 7.4

to this theory, we can think of the universe as in some sense starting out with just one branch; then as things begin to happen at the microscopic level it splits and splits into countless numbers of alternative parallel branches, each of them very slightly different. This means that everything that *can* happen *does* happen somewhere. But the price you have to pay is that we, the observers, also split and have to inhabit all these many different universes. So, for example, in the cat experiment we very neatly explain the paradox. Is the cat alive, is it dead or is it both together? Well, the answer is that when you perform the experiment, the world is split into two, with one having a live cat and its observers and the other having a dead cat and its observers. Hence there is this branching of the tree-like structure and the universe is constantly multiplying over and over again.

There is obviously a time asymmetry here, inasmuch as we are talking about the universe *splitting*. We could envisage the time-reverse where the branches fuse, or alternatively the splitting takes place backwards in time. It then becomes interesting to discuss why it should be that the universe is splitting and not fusing. In the many universes interpretation no collapse takes place when you make a measurement or observation. All that happens is that the observers go off into different branches, which they *interpret* as a particular world having been selected. However, we have not really selected a particular world since what has actually occurred is that our minds have been split into two or more and each one has gone off into a different branch. From the point of view of any given mind going up the branches of the tree, it would appear that the world has been split and that there has been some irreversible jump. But when we look at the system overall we see that that isn't the case. When we write down the wave function there is no preferred time direction. So we must suppose that there is also this fusing set (fig. 7.4) and that there are as many worlds in which things are, crudely speaking, going backwards in time and in which these measurement processes are also taking place backwards in time. And of course those observers think everything is perfectly normal because, relative to what they are doing, the flow of other physical processes is all in the right direction.

The foregoing seems to suggest — at least to me, and this is a very personal interpretation of this many universes theory — that the direction of the branching is inextricably tied up with the *arrow of time* (if I may use that expression), associated with the world which the experimenter inhabits. For example, if we wanted to carry out the cat experiment, we would have to be in a world in which the cat was alive before the experiment. And if we were doing it in the other situation we would, according to that convention, have to be in a world where the cat was alive after the experiment. This is perhaps just a rather complicated way of saying that the splitting direction, the irreversibility in the quantum measurement, follows on behind the arrow of time which is imposed by the thermodynamics of the world in which the observer finds himself.

I now want briefly to mention two other points of view. The first is called the *statistical interpretation* of quantum mechanics, which I don't really want to get into because it forsakes any attempt to describe what is actually going on in the world. According to the statistical interpretation, we can never say what any *individual* electron, or whatever, is doing; we can only say what happens to *collections* of such things. The second approach I wish to discuss briefly is the *hidden variables* approach. Hidden variable theories are attempts to find a classical face behind the quantum face of the universe. These are an attempt to return to classical physics and to abandon the weird overtones injected by quantum physics. They seek to demonstrate that the uncertainty or fuzziness we have been discussing is really of the type associated with the weather, the stock market and so on, and is essentially of a classical nature.

According to the hidden variable theories, the wave function does not give a complete description of the state of a physical system. For it leaves out of account additional 'hidden' attributes of the system — attributes which, together with the wave function, jointly determine what result an observer will get when he carries out a measurement. The measurement then in principle serves to reveal the state that the system was in before the measurement was carried out. On this view, it is only our prior ignorance of the exact state of the hidden variables, something which is left out of account in the wave-functional description, that prevents our saying with certainty what result the measurement will yield.

Now it is possible to prepare pairs of systems, particles say, with correlated states. For example, one can generate pairs of photons with correlated polarizations. This means that if you place identical polarizing filters, set at the same angle, in the paths of the two photons, with photon detectors behind them, then either both will pass through and trigger the detectors or neither will, whatever angle you choose. According to a hidden variable theory, these two photons, even before they encountered the two filters, are in a determinate state of polarization from which, were one able to know it, one could predict with certainty, for a polarizing filter set at any specified angle, whether or not the photon would pass through.

In 1964, however, John Bell demonstrated that in order to reproduce the statistical predictions of standard quantum mechanics, these hidden variables, in the case of such correlated systems, would have to be *non-local*. Unless the two particles were somehow able to communicate with each other, so that a change in the hidden variables for one of them brought about a change in the corresponding hidden variables for the other, then the correlations one would obtain when carrying out measurements on the two particles, would, if such a hidden variable theory were true, be different from what quantum theory predicts. Imagine that one is repeatedly shooting out pairs of photons with correlated polarizations and passing them through polarizing filters with photon detectors behind them. And suppose further that one is allowed independently to vary at random the orientations of the polarizers that will eventually be encountered. Suppose, finally, that a hidden variable theory is true and there is no way in which one particle, or the orientation of the corresponding polarizer, can directly affect the other particle. It follows from Bell's work that a certain mathematical relationship, known as *Bell's inequality*, must then be satisfied. This states that the correlations one obtains, between the results obtained from the spatially separated detectors — that is, the degree to which detection or non-detection of one photon will be correlated with detection or non-detection of the other, for different settings of the polarizer orientations — cannot, when mathematically combined in a certain way, exceed a certain value. And for certain angles between the filters, quantum mechanics predicts, to the contrary, that this value *will* be exceeded.

Of course that in itself doesn't show that local hidden variable theories are incorrect. For it could be that this particular prediction of quantum mechanics is mistaken. It is something that must be tested experimentally. In 1982 the French physicist Alain Aspect put Bell's inequality to the test in the most crucial situation — that in which the two separated systems (photons and detectors) would not only have to communicate with each other, in order to produce the kind of correlations that quantum mechanics predicts, but would have to do so *faster than the speed of light*. In Aspect's experiment, the orientation of the polarizer a given photon would be directed towards was controlled by a rapid switching device, which was able to determine the orientation of the polarizer that each photon would encounter while it was actually in flight. The result, as most physicists expected, was wholly consistent with conventional quantum mechanics, producing a clear violation of Bell's inequality. In the light of Aspect's experiment, it looks, therefore, as though one can continue to hold a hidden variable theory only if one is prepared to countenance something like faster-than-light signalling, which most physicists would regard as too high a price to pay.

I'm going to finish my discussion by focussing on what I call the *Bohr—Wheeler* interpretation. Wheeler has, in fact, very recently brought out a most important connection between the act of measurement in quantum mechanics and the nature of time. This is based on his so-called *delayed choice* experiment which is similar to the Young's two slit experiment. Let me briefly describe it. Referring to fig. 7.5, we have a source of light on the left-hand side. S is a screen with two apertures which incorporates a lens. I is the image screen which is also equipped with a lens. This image screen is actually in the form of a venetian blind, so we can open the screen and let the light through. Finally, on the right hand side, we have two photon detectors. If the venetian blind is open, any light passing through the upper slit of screen S is focussed by the lenses to arrive at the lower photon detector, while light passing through the lower slit arrives at the upper detector.

Now the point is this. As I said at the outset, the Young's two slit experiment exposes two complementary descriptions of the world and not two contradictory descriptions. You can perform the original Young's experiment, in which case you don't enquire as to which slit the photons have gone through, and an interference pattern is observed. We achieve this by keeping the venetian blind closed, and it then displays an interference pattern. However, if we choose to open the venetian blind and allow the photons to pass through the screen I, the detectors will tell us from which slit the photons came. So if the bottom detector goes click we know the photon will have come via the upper route and if the other one goes click we know the photon has taken the lower route. The point is that the experimenter can clearly wait until after the photon has passed through the slit system before deciding whether or not to open the venetian blinds. In other words, what the experimenter can do, by making a decision at a late stage, is to determine whether or not our description of reality in the past is such that we can say the photon did take a particular path, upper or lower, or that our description of reality is

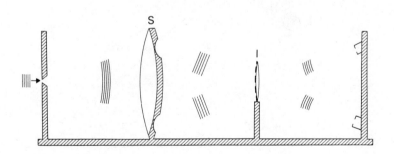

Fig. 7.5 *Delayed choice*

such that it in some sense took both paths or that both these worlds continue to coexist and overlap.

We have here a most peculiar relation to time because there is a sort of retroaction. It is important to realize that you cannot actually use this device to signal into the past, since there is no way you can manipulate the venetian blind to send a signal backwards to the source of the photons. This is because there is always a degree of unpredictability, in that you can never know in advance of the measurement whether it was the upper or the lower route that the photon took. However, the particular aspect of reality (measuring the route and seeing the particle-like aspect, or not measuring the route and seeing the wave-like aspect) you wish to reveal depends (or can depend) on the later choice of the experimenter.

Again you might say, well how long does a photon take to go across the laboratory? A few nanoseconds perhaps; and who cares if there is retroactive causation on such a small time scale as this? Wheeler has pointed out, however, that the Great Architect has provided in the universe something very much like this two slit system. There are a number of quasars, distant objects billions of light years away — which are seen in multiple image form. The reason for this is illustrated in fig. 7.6. Q is the quasar, G is a galaxy and O is the observer. The intervening galaxy, G, is able to bend the light — gravity of course bends light — and the gravity from this galaxy can bend the light in either way much like that in Young's slit experiment. We can now envisage bringing the light beams from both paths together, overlapping them and performing an interference

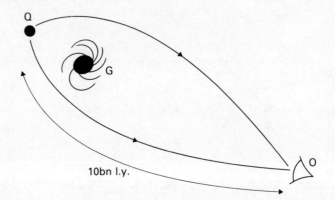

Fig. 7.6

experiment. Carrying out this interference experiment is of course something else because in general those two light paths will differ by quite a considerable amount. But one can imagine that if you could move around the universe at will and set everything up in just the right way, then in principle it would be possible to set up this interference experiment. The arms of the apparatus are now billions of light years long and the look-back time, or the backwards causation time, is now an appreciable fraction of the age of the universe.

Wheeler has pushed this whole business to the extreme by pointing out that it is possible, when we are looking at distant astronomical sources, to affect by the observations that we choose to make *now* the reality of what *was* — even at a time when there were no observers in the universe! In fig. 7.7, a famous drawing is reproduced which is supposed to summarize this sentiment. The top left of the U symbolizes the beginning of the universe, which then develops and evolves. Eventually intelligent creatures arise (symbolized by the

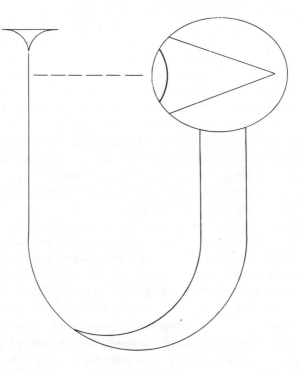

Fig. 7.7.

eye) and look back, in the way I have just described, to the very early stages of the universe and in some sense have a hand in congealing the early universe into concrete reality. You can think of this as a self-excited loop or a feedback circuit where you can't really separate the reality of what was from the observers who will be. Of course, 'the observers who will be' depend on the universe giving rise to them in the first place; and so the whole thing is coupled together in a holistic and mystical fashion. This is Wheeler's point of view. It is, in a sense, the Copenhagen view taken to the logical (and, some might say, ludicrous) extreme.

The problem with the official view and the reason why people are forced into such bizarre ideas as the many universes interpretation is simply that when one comes to apply quantum physics to the universe as a whole one is faced with a fundamental problem. The problem is that if everything you are trying to describe belongs to the quantum system, who is going to be observing it? Who is going to collapse the wave function? These various positions which I have been outlining all try to address that issue in one way or another.

I want to finish my discussion by mentioning some ideas of my own which are very fragmentary and may not be at all original. In effect, I shall be asking questions, rather than solving problems.

I am convinced that the problem of quantum measurement is intimately associated with the mind—body problem. I do not want to introduce mind-over-matter, as Wigner would have us do; but I believe it's very important when we are talking about the mind—body problem to distinguish between different categories or different levels of description. When we talk about the mind—body problem we have to accept that the mind belongs to a different level of description from the brain, and that mind and brain represent *complementary* descriptions of the same system and not contradictory ones. I like to draw an analogy between hardware and software in computing — the brain is like the hardware and the mind is rather like the software. We associate mind with information, input and output, processing and so on, while we associate the brain with little electrical circuits and so forth. In my view the wave—particle duality in quantum physics is not only very closely analogous to these other dualities but is actually part of the same basic problem. The reason for this is that the quantum wave encodes the *information* that we have about the system. So that's a bit like the software. On the other hand the particle-like aspect is rather like the hardware. Therefore the act of

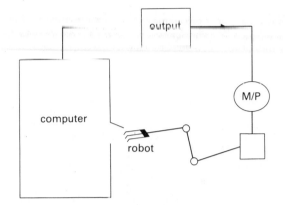

Fig. 7.8

making a measurement means that we are really coupling these issues of mind—body and particle—wave together.

The problem arises in quantum physics because it is very hard to separate out the hardware from the software. The point is that if we think of the hardware as the particles — the electrons and so on — then I would say they 'support' the software — the wave function — because the form of the wave function depends on what the particles are doing. On the other hand the behaviour of the particles is controlled by the wave function. For example, the wave function changes abruptly when we make a measurement, and therefore what the particle can subsequently do also changes abruptly. So we seem to have an entanglement between the hardware and the software in which each one has a hand in determining the other, and I think that the famous quantum measurement problem arises when we artificially try to untangle them. I have tried to think of some sort of analogy of hardware entangling the software but it is very difficult, and fig. 7.8 is just a simplified attempt. Imagine a computer in which the output is coupled to a microprocessor which in turn controls a robot arm that can dip inside the computer and rearrange the circuitry. Then you might think of this as a sort of loop enfolding hardware and software together. Perhaps this is the sort of direction in which we should be thinking.

By way of summary, let me refer to fig. 7.9. We can think of the particle activity as giving rise to effects in our apparatus and that out there somewhere is the observer who looks at the apparatus. We

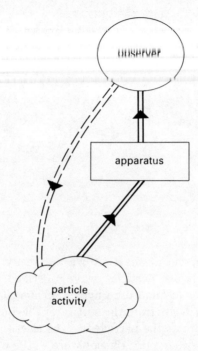

Fig. 7.9

have a logical chain coupling the atomic level through the apparatus to the observer. However, in some way the observer is coupling back to the particle activity. Indeed, as we have seen in Wheeler's delayed choice experiment, this particle loop can actually stretch backwards in time. Not, to be sure, in a way that allows us to signal into the past, but in a way which, nevertheless, challenges our ordinary notions of time. I would see the hardware—software entanglement as a little bit like this loop.

I will end by going back to my starting point, which was Bohr. Bohr held that the act of quantum measurement is complete when the result can be communicated between people in plain language or when the results of the observation have some sort of meaning for us. In this loop that I have been talking about, we start out with hardware (the electrons say) and end up, through some process which is still only dimly understood, with information and meaning. Finally, we loop back round again to couple with the hardware.

I don't think that I have answered very much. What I have tried to do is conjure up certain images and suggest a new way of looking at this problem. If I have succeeded in nothing else, I hope at least that I have managed to stimulate some thinking on this perplexing subject.

FURTHER READING

Many introductions to the essential ideas of quantum mechanics can be found in Davies, P. C. W. and Brown, J. R., *The Ghost in the Atom* Cambridge University Press, Cambridge, 1986), which also includes a discussion of Aspect's experiment, and Rae, A. I. M., *Quantum Physics: Illusion or Reality* (Cambridge University Press, Cambridge, 1986), which gives a very clear derivation of Bell's inequality.

For a more advanced treatment, with greater emphasis on the philosophical issues, see Gisenbud, L., *Conceptual Foundations of Quantum Mechanics* (Van Nostrand Reinhold, Wokingham, 1972) and d'Espagnat, B., *In Search of Reality* (Springer-Verlag, Berlin, 1983).

The many-universes theory is developed in detail in De Witt, B. S. and Graham, N., *The Many-Worlds Interpretation of Quantum Mechanics* (Princeton University Press, Princeton, 1973).

A proposal to introduce consciousness directly into quantum mechanics is given by Eugene Wigner in *The Scientist Speculates*, ed. I. J. Good (Heinemann, London, 1962).

General surveys of time asymmetry in physics, with some attention given to the asymmetry involved in the collapse of the wave function, include Davies, P. C. W., *The Physics of Time Asymmetry* (Surrey University Press/California University Press, Berkeley, 1974), Penrose, R., 'Singularities and time asymmetry', in *General Relativity: An Einstein Centenary Survey*, ed. S. W. Hawking and W. Israel (Cambridge University Press, Cambridge, 1979), and Prigogine, I., *From Being to Becoming: Time and Complexity in the Physical Sciences* (W. H. Freeman, Oxford, 1980).

A discussion of the delayed-choice experiment and other fundamental topics can be found in Wheeler, John, 'Beyond the black hole' in *Some Strangeness in the Proportion: A Centennial Symposium to Celebrate the Achievements of Albert Einstein*, ed. H. Woolf (Addison-Wesley, Reading, Massachusetts, 1980).

The essentially non-local character of quantum mechanics has been strongly emphasized by David Bohm in *Wholeness and the Implicate Order* (Routledge and Kegan Paul, London, 1980).

The hardware—software relationship and the importance of keeping a distinction between different levels of description is beautifully developed in Hofstadter, D. R., *Gödel, Escher, Bach: An Eternal Golden Braid*

(Penguin, Harmondsworth, 1980). See also Hofstadter, D. R. and Dennett, D. C. (eds), *The Mind's I* (Penguin, Harmondsworth, 1982).

A discussion of the analogy between our perception of the flow of time and our perception of dizziness is given in chapter 9 of Davies, Paul, *God and the New Physics* (Dent, London, 1983)

8

The Open Future

J. R. LUCAS

In the previous lectures of this course we have been considering time from the physicists' point of view. By limiting the questions to those the physicists ask, we have been able to go a long way towards obtaining clear and definite answers. But a price had to be paid. We could not ask many of the questions we wanted to ask, and the characterization of time given was only a partial characterization. The time of the special and general theories of relativity is, like space, directionless. There is no natural direction of time in Newtonian mechanics, in electromagnetism, or in the field equations of general relativity. The big bang could, quite possible, be paralleled by the big crunch. In other branches of physics there is, however, a temporal asymmetry. The Second Law of Thermodynamics has been expounded by Dr Atkins in chapter 6, and we have a further account by Professor Davies of the other temporal asymmetries in physics in chapter 7. Time thus far characterized has two of the features we recognize from our own experience: it is some sort of dimension, and it has a direction; but it still lacks some other typical features, in particular the modal differences between future and past and the uniqueness of the present. I can bring this point out by referring to fig. 6.1 (p. 81), which Dr Atkins used when he was plotting the entropy of the universe throughout its whole existence, from its beginning at A, to its end at B. I thought (when I was listening to his lecture, since he is sometimes thought of as a militant atheist who has rewritten the book of Genesis, that when he turned his attention to the last book of the Bible and rewrote the Apocalypse, he might revert to the Greek alphabet, and have A and Ω instead of A and B to name the beginning and the end.) He also indicated a point C, which was where we were, more or less it did not matter, a thousand million years either way. But although it does not matter a thousand million years either way to the physicist interested in the Second Law of Thermodynamics — entropy was increasing in the Pre-Cambrian age and is likely to go on increasing in a giga-year's time — it does matter like anything to us whether we are talking of the past, the present, or the future. A nuclear winter at the end of the Jurassic age is — so to speak — water under the bridge. Of course, I am awfully sorry for all those poor old dinosaurs, but life has always been a chancy business, and actually the big freeze set the scene for the emergence of the warm-blooded mammals and birds. A nuclear winter at the end of this century is, however, a distinctly chilling prospect, and although we are not so immediately concerned about the ultimate heat death of the universe that Dr Shallis was talking about, it too has worried many men thinking in the long term what future there was for mankind.

Our different attitudes towards the future and the past are due, in part, to our being able, as we suppose, to make a difference to what happens in the future, but not to what has happened in the past. It is no good crying over spilt milk, whereas the future is something we can try to do something about. Hence as agents we are committed to viewing the future quite differently from the past. The future is open, alterable, to some extent malleable by us. The past is closed, unalterable, part of the irrevocable record of history. And the present is the link between the two, the time of decision, when we have to come down on the one side or the other, and the moment of truth, when hopes are finally shown up as dupes or fears as liars.

The agent's concept of time is modal. Time is the passage from possibility through actuality to necessity. It clearly has a direction, and to that extent coheres with what some physicists are impelled to say, but it clearly goes much further than physicists want to go, both in the uniqueness ascribed to the present and in the different status ascribed to the future and the past: and it is often maintained that the agent's concept of time is no longer tenable in view of the discoveries of modern physics, and must be relegated to the museum of outworn ideas as a prime example of outdated anthropomorphism.

Now, actually, I don't take the charge of anthropomorphism as being so serious an accusation as it is often supposed to be. I am a man. I make no bones about it. Although it is quite right for a scientist in his professional capacity — when he is wearing his white coat, so to speak — to distance himself from a number of adventitious human concerns and assumptions, it is quite wrong, and a serious breach of intellectual integrity, for him to put off his humanity altogether, and to make out that because he does not know something as a scientist, he does not know it at all. In our generation scientists have come to recognize their moral responsibilities, and appreciate that although a particular science cannot answer questions of right and wrong, that does not mean that these are questions that scientists can put on one side. In the same way, it would be irrational when thinking about time to put on one side the knowledge gleaned from our own experience of time, either as observers or as agents. We could not be scientists unless we were also agents. There might conceivably be Martians or non-human androids from another planet in another solar system who carried out experiments, made observations, and were able thus to arrive at well-founded scientific theories; but only because they *carried out* experiments and *made* observations — that is to say, only because they were agents. Agency is not only a fundamental fact of human existence, but is presupposed by even the most speculative natural science. It is, indeed, the starting point for all philosophical inquiry. Instead of Descartes' *Cogito, ergo sum* — I think, therefore I am — we should take our stand on the even more intuitive principle, *Ego, ergo ago*, I am, therefore I act. And to act is to decide, and to decide is to choose between alternatives, and to select one and actualize it, and reject the others as unrealized possibilities. This view of time, although intuitive and widely accepted in ancient and medieval thought, has always been contested. Some thinkers have denied the necessity of the past, and held that the past is no more incapable of being influenced than is the future. Michael Dummett, in chapter 9, argues that some standard objections to the startling idea that what we now do can influence the past are philosophically unsound. Without here attempting to rebut Dummett's arguments, I shall assume in what follows that the commonsense view of the past as fixed is correct.

What I want to deal with here is the much more widely held opinion that rather than the past being alterable like the future, the future is, contrary to appearances, unalterable like the past. It has been argued for on many grounds: on logical grounds from the timeless nature of truth; on theological grounds from the omniscience of God; on scientific grounds from the predictability of classical physics; and, most relevantly to this course of lectures, on relativistic

128 The Open Future

grounds from the nature of time as a fourth dimension and the way in which simultaneity is relative to one's frame of reference.

The logical argument from the nature of truth was first discussed by Aristotle, in chapter 9 of the *De Interpretatione*. Aristotle considers a prediction — his example is that there is going to be a sea battle tomorrow — and asks whether it is true or false. Of course, we don't know now whether it is true or false. But in the fulness of time we shall know that either in the event it proved true or else that in the event it proved false. But if it is true then, it is true now, and if it is false then it is false now. So as of now it is either true — in which case the sea battle cannot not happen tomorrow — or else it is false — in which case there cannot be a sea battle tomorrow. So the die is already cast, and it is not in our power to choose otherwise about what we do tomorrow than in accordance with the already existing truth.

Aristotle is sure that there is something wrong with this argument, as I am sure you are. The difficulty is to say precisely what. Almost any solution that has been offered is open to some fatal objection: and almost every solution has been offered — the passage has been more argued about and commented on than any other in Western philosophy — I once began to make a bibliography: it was easy to begin with Abelard and Ackrill, but as I went my way through the As, with Anscombe and Anselm, Aquinas and Aristotle, Augustine, Avicenna and Averroes, my spirits began to flag, and I knew I should never get through the Bs, with Boethius, Bradwardine and Broad, let alone get through the whole alphabet. And the power of the argument to confuse even the clear-minded is shown by three cases which turned up in the Court of Appeal and the House of Lords about 12 years ago. It was held that section 14(1) of the Trades Descriptions Act, 1968, '. . . has no application to statements which amount to a promise with respect to the future, and which therefore at the time when they are made cannot have the character of being either true or false . . .'

I am not going to try to unravel all the arguments that go to create Aristotle's problem about tomorrow's sea battle — I have fared no better in my attempts than my medieval predecessors. I shall be equally cursory in my treatment of divine omniscience, which seems to me to turn partly on a muddle about what it is to know something, partly on a mistaken and unChristian view of God. In an ordinary way of speaking we can perfectly well foreknow the future without foreclosing the freedom of agents. People knew in advance that I was going to give the lecture on which this chapter is based. But that did not mean that I had to lecture, or lacked any freedom of the will. I *could* have not lectured. If, shortly beforehand, I had been persuaded by one of my pupils to embrace Existentialism, and decided to manifest my freedom by committing an *acte gratuit*, and instead of giving the lecture, had popped off to Paris to attend a way-out art exhibition on the Left Bank, I could have done so. It would have been hard on the audience, of course. Instead of saying, as they were in fact safely able to, that they knew I should be there, they would have had to say that they only thought they knew. Knowledge is, thus, subject to a retrospective withdrawal proviso. So long as the prediction works out, you really did know all along: but if the prediction proves false in the event, then the knowledge claim has to be withdrawn, and you never really knew what you thought you did. This causes no great problems when we are thinking of human knowledge — it only goes to show that humans are fallible, which we knew all along. It is different, however, with God. Could God make a mistake? Boethius thought the very suggestion blasphemous as he pondered freedom and fate in the last book of his *Consolations of Philosophy*, and so was led to propose far more difficult resolutions of the problem. But is it blasphemous? It seems entirely counter to the Greek idea of God, who was the altogether Mostest, the Unmoved Mover, absorbed in contemplation of His own perfection, but it seems to me — although the Master of Balliol controverts it — entirely consonant with the Christian view of God that He should expose His predictions to the risk of falsification, as He does His plans to the risk of frustration, by human fickleness for the sake of human freedom.

The scientific argument for determinism used to be very strong but now has lost almost all its cogency in view of the indeterminist nature of quantum mechanics. Although in the nature of the case we cannot give an absolutely watertight proof that there could not be some 'hidden variable' theory that was determinist and yet yielded the same results as quantum mechanics, there have been a whole series of arguments from those of von Neumann onwards which make it increasingly implausible to suppose that quantum mechanics could be anything other than probabilistic, and the experimental results of Aspect in Paris in 1983 have all but clinched the issue. So far as quantum mechanics is concerned there is no argument for determinism, and, although as we shall see there is still more to be said on its bearing on the openness of the future, it does not at all support the view that the future course of events is already fixed and unalterable, like the past.

Relativity regards time as a dimension, like, although not entirely like, the dimensions of space. It encourages us to treat space—time geometrically, and in general relativity to fuse geometry and physics in a single geometrodynamics. It is an inevitable concomitant of this approach that we take a block view of the universe in which the future course of events is already laid out as a path in Minkowski space—time. The future already exists: it is only that we do not yet know what is in store for us, and only discover that as we crawl along our world-line. More cogently, it argues that there cannot be a real, modal or ontological, difference between future and past, because the present — the instant that divides past from future — depends on our criterion of simultaneity, and our criterion of simultaneity depends on our frame of reference. If we change our velocity, we change our frame of reference and our criterion of simultaneity. We change not only what distant events we take to be simultaneous with us — i.e. what we take to be present — but concomitantly what distant events we take to be past or future. If I were to ascribe a date to a distant event (say the arrival at α-Centauri of a space rocket dispatched aeons ago) I might, if I had the A frame of reference (shown by $t_A = 0$ in fig. 8.1), regard it as still being in the future: but if I were going at a great velocity towards it, I should correspondingly counter-incline my space axis, indicating what events I should regard as simultaneous with me, and then regard that same event as in the past — the B frame of reference (shown by $t_B = 0$ in fig. 8.1). Whether it is future or past depends entirely on my frame of reference, and thus cannot be anything about the event itself but is entirely relative to me, and my temporal perspective on it.

This argument seems quite telling, and for some scientists seems to pose an excellent example of a confrontation between scientific truth and philosophical prejudice. But it is not really like that. For one thing it is a conflict not only between physics and philosophy but between one branch of physics and another. If we consider only relativistic physics, the world of black holes and temporal horizons that was revealed to us in chapter 2, we are led to think in terms of a block universe in which the future is closed, but if we consider quantum mechanics in which wave functions collapse into particles, or operators when measured assume *eigen*-values with a probability given by their projections on the corresponding *eigen*-vectors, no suggestion of the future's being closed can be entertained. Indeed,

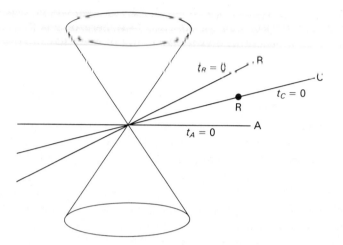

Fig. 8.1 *R is rocket's arrival at α-Centauri, lines A, B and C denote respectively the simultaneity axes of frames A, B and C*

not only is the future not closed but the past is in danger of being left open. You remember Schrödinger's cat, which according to orthodox quantum mechanics does not really die until someone looks inside its cage to see how it is getting on. It reminds me of the tree in the quad which, according to the phenomenalists, is not there unless someone is looking at it. Schrödinger has out-Berkeleyed Berkeley. Berkeley was talking only of things existing, and saying that for something to exist was for it to be perceived: but Schrödinger extends this to events taking place, and seems to be saying not just *esse est percipi*, to be is to be perceived, but *genesthai esti horathēnai*, to happen is to be perceived.

There are obviously great literary possibilities — the cat that may not be looked at by the Queen for fear of the royal glance proving fatal, or, to change the roles, the Princess who has been turned by the sorcerer into a superposition of dead and alive states, and the brave hero who dares to take a chance and gaze upon her, is rewarded for his bravery by her collapsing into being alive and in his arms. But scientists are unhappy at the prospect in spite of its science-fiction possibilities, and Professor Penrose posited an effect of a gravitational field to secure the collapse of the wave function at each present juncture rather than wait in suspended possibility until the intervention of an observer constituted the moment of truth (see chapter 4). Professor Penrose saw this as a merit of this theory, and whatever view other physicists take of his theory, they are unhappy with unactualized possibilities in the present and past, and try, one way or another, sometimes even — as Professor Davies did once — positing split universes, to accommodate the openness towards the future suggested by quantum mechanics without compromising the actuality of the present or the unalterability of the past.

Whatever account we give of the collapse of the wave function, the relativistic account of simultaneity is faced with a problem not only of human choices but quantum-mechanical events. If there is an ontological difference between future and past, then there is a real difference throughout the universe, giving us an absolute concept of simultaneity not dependent on the frame of reference, contrary to the teaching of the special theory of relativity, as normally understood. But the real meaning of the special theory of relativity is often misunderstood. There is a confusion between what frames are equivalent to one another in respect of *certain physical* laws and what frames are equivalent to one another in *all* respects. The best way of bringing this out is to follow Mr Newton-Smith's example in chapter 3 and to discuss the parallel case of classical mechanics. Classical mechanics is, as Mr Newton-Smith pointed out, relativistic not only with regard to displacement and re-orientation but with regard to velocity as well. There is no way, so far as the laws of Newtonian mechanics go, of telling whether a frame of reference is at absolute rest or moving with a uniform absolute velocity. Newton was quite clear on this point. But although he could not tell *by means of Newtonian mechanics* whether a frame of reference was at rest or not, it did not follow that it was a meaningless question, or one that he could not possibly have answered. He was inclined to think, although he could not be sure, that the centre of mass of the solar system determined such a frame of reference, and he might have found a more certain answer in his theological researches — he might have found it revealed to him from his study of the Book of Daniel. Equally, at the end of the last century, if the result of the Michelson—Morley experiment had been positive, we should have been able to determine the velocity of the Earth relative to the ether, and the ether itself would have constituted such a frame of reference at absolute rest. Although Newtonian mechanics could not by itself single out any one of a set of inertial frames of reference as being at absolute rest, Newtonian mechanics plus electromagnetic theory might have been able to.

In the same way, although the special theory of relativity cannot single out any one of a set of frames of reference that are equivalent to one another under the Lorentz transformation as being at absolute rest, there is no reason why the special theory *plus* some other theory or some other consideration should not do so. This indeed happens in some versions of the general theory of relativity, where we sometimes distinguish a preferred frame of reference which we regard as being at rest. And we might have reason to do the same if we adjoined to the special theory some version of quantum mechanics, or supposed our special relativistic universe peopled by rational agents. To be precise, we might think of some interaction between a microphysical system on board our space rocket and the microphysical system of α-Centauri which would take place as soon as the rocket arrived, and then ask how this date tallied with our two frames of reference, A and B (fig. 8.1). It is unlikely that we could tell, but it would no longer be in principle meaningless to ask the question. Let us consider the supposition that the collapse of the wave function is, in fact, simultaneous with our raising the question on Earth. What follows? As far as our relativistic physics is concerned, nothing follows. The event is outside the light cone in the 'absolutely elsewhere'. It cannot affect, or be affected by, anything we do here and now. But we shall be able to distinguish between frames of reference when we are concerned not with electromagnetic questions of physics but modal questions about time. Neither the A nor the B frame of reference will be apt for that, but instead a third one, the C frame (fig. 8.1). That will define the real relation of simultaneity and the corresponding time axis — the state of absolute rest. And this will be no worse than what already we are familiar with in general relativity, and what Newton envisaged for Newtonian mechanics.

It follows from this, I think, that there is no incompatibility between the concept of time appropriate to agents deciding their own future course of action, and so influencing the future course of events, and the concept of time put forward by physicists. For some purposes — perfectly legitimate although not at all obligatory — the physicists need to consider only certain aspects of time and to neglect others: in so doing they are doing nothing wrong; but the fact that they are choosing to ignore certain features does not mean that we must ignore them too when we are considering other aspects of life, or indeed other aspects of physics. Nor is the openness of the future incompatible with the other time-directed feature of physics to which Dr Atkins gave eloquent expression. The Second Law of Thermodynamics teaches us that entropy increases, and that we must inevitably have some experience of 'chaos as we sink into equilibrium and the grave', but the openness of the future assures us that at least we have some measure of choice over the precise path we follow in the years that are left to us before we finally die.

FURTHER READING

The first discussion of the open future is given by Aristotle in chapter 9 of his *De Interpretatione*. The best translation into English is by J. L. Ackrill, (Oxford University Press, Oxford, 1963). It has useful notes. The *De Interpretatione* was much discussed in the Middle Ages and again in recent years. A useful account of the issues from a modern point of view (though with some errors in the formal derivations) is given in 'Truth and Necessity in Temporal Perspective' by N. Rescher in R. M. Gale, ed., *The Philosophy of Time*, (Macmillan, London, 1968).

White, Alan R.., *Truth*, (Macmillan, London, 1970), ch. 3, pp. 141—56, has a good discussion of future contingents from the point of view of ordinary language and common sense. A more polemical account is given by Gilbert Ryle in chapter 2 of his *Dilemmas*, (Cambridge University Press, Cambridge, 1954). The suggestion that special relativity entails determinism is put forward by Hilary Putnam in 'Time and Physical Geometry', reprinted in his *Philosophical Papers*, (Cambridge University Press, Cambridge, 1975), and refuted by Richard Sorabji in his *Necessity, Cause and Blame*, (Duckworth, London, 1980, pp. 114—15). See also Storrs McCall, 'Temporal Flux', *American Philosophical Quarterly*, 3 (1966), pp. 270—81.

Ω

Causal Loops

M. DUMMETT

I

I was originally asked by the organizer of this series to give a talk about the direction of causality. After I had accepted, he later rang me up and asked me to put something in about time travel, as it figures in science-fiction stories; so I gained the impression that my talk was intended to offer a little light entertainment to round off the series.

Let us begin, then, with what I presume is the first of the time-travel stories, H. G. Wells' *Time Machine*. Most of the book is taken up with the time traveller's experiences in the very remote future; but it opens with a meeting between him and a friend to whom he tries to explain his invention of a device for travelling in time. Time, he says, is just another dimension, like the spatial ones; after all, what lacked all duration simply would not exist. He then demonstrates time travel with a miniature time machine that he has constructed, in all but size just like that in which he proposes to travel. He sets it on the table, adjusts the controls for it to travel into the future, and switches it on: its appearance becomes blurred for a second, and then it vanishes. Where has it gone? Into the future. Since it has no occupant to bring it to a halt or reverse it, it will never be seen again.

Suppose you are the time traveller's friend, and that this has just happened. The time machine has incontestably disappeared: but has it really gone into the future? It is 6 p.m. on Friday. If the time machine has gone to tomorrow, Saturday, even though only passing through, it should be there tomorrow. But it is not, and the time traveller does not expect it to be. What, then, does it *mean* to say that it travelled into the future, if it is not there when the future comes? Of course, the reason why we are supposed to be unsurprised at not finding it there tomorrow is that we think that, by then, it is far ahead of us. We are playing the tortoise to the time machine's Achilles (in a race in which they start from the same point). When the tortoise reaches the 100-yard post, he should not expect to see Achilles there: he passed it long ago, and is at the 1000-yard post by then. The picture to which Wells is surreptitiously appealing is that of a fourth dimension along which we are already travelling at a certain rate, namely an hour per hour: the time machine simply goes much faster, say at a day per hour. On this picture, any given location along this fourth dimension will be in a certain state at any given time, say *now*: so it will make sense to say, at 7 p.m. on Friday, that the model time machine has *now* reached 6 p.m. on Saturday, and, at 6 p.m. on Saturday, that it is *now* more than three weeks ahead. Thus at 7 p.m. on Friday — that is to say, when *we* are at 7 p.m. on Friday — it is true of 6 p.m. on Saturday that there is then a model time machine on the table; but, of course, by the time we have reached 6 p.m. on Saturday, it will no longer be true of that time — that location in the fourth dimension — that any such time machine is at that, or any other, place. I need not waste your time exposing the incoherence of this picture. A writer of popular books on time, J. W. Dunne, gained a reputation in the thirties by taking quite seriously the very picture to which Wells was appealing. Observing that time travel involves two kinds of time, that along which the journey is made and that which the journey takes, he thought that the argument must be able to be reiterated, arriving at the conclusion that there must be infinitely many temporal dimensions. It seems at first sight, therefore, that the whole idea of time travel rests on a primitive misrepresentation.

Yet it seems to make sense when we introduce a traveller: is it, then, consciousness that makes the difference? Wells assumes that travel in time is like travel in space, in that, in order to reach a given time, you have to pass through the intervening times. If, then, we agree that, for something which remains in a particular place to pass through any given time, it must be in that place at that time, we must revise our picture of the time traveller's journey into the future. There will be no flickering and disappearing on the part of the time machine: it must remain where it is throughout the interval between his getting into it and his getting out of it in the remote future. The only special feature is that time will pass with immense rapidity for him: at the remote future date at which he switches off the machine, he will have aged only a week or so. It is not just an effect on his consciousness, therefore: all his bodily processes will have been tremendously slowed down. This, from the standpoint of everything else, is the effect of the time machine. Nothing appears unusual about it (save possibly its resistance over long periods to decay): it merely slows down any process taking place within it.

We can now revise the story of the model time machine. Let us suppose placed in it a small watch — one of those that show the day of the week as well as the time. When the machine is set going, nothing spectacular happens: it remains where it is. It is only that the watch appears to have stopped: by 6 p.m. on Saturday, it is reading only '7 p.m., Friday'.

This may perhaps deserve to be called a 'time machine', but it hardly appears to be a vehicle for *travelling* in time in any but a highly subjective sense. A quite different interpretation is obtained if we allow a jump to occur: if it is set to advance 24 hours, the machine, when started at 6 p.m. on Friday, immediately vanishes; it as suddenly reappears in the same place at 6 p.m. on Saturday, the watch it carries being exactly one day slow. Here it is not only what the machine carries that behaves unusually, but the machine itself in relation to its environment, involving a gross discontinuity; possibly the most surprising thing is the displacement of the air molecules in anticipation of its reappearance.

An analogous interpretation of a backward move in time carries greater difficulties. Suppose that the time traveller sets the model machine to jump to 24 hours earlier. Then at 6 p.m. on Thursday it must suddenly have appeared on the table: the time traveller and his friend might be able to verify this if a movie camera had been trained on the room since before that time. What, then, happens at 6 p.m. on Friday? There are two cases. In the first, the time traveller starts the machine off, whereupon it promptly disappears. His friend exclaims, 'Where's it gone?', looks around, and spots it on the mantlepiece: on examining it, they find that the watch reads 5 past 6, Saturday. The explanation is that, on Friday morning, the cleaning lady came in and removed the machine to the mantlepiece in order to polish the table. As a variation on this, we may suppose that, when the time traveller and his friend entered the room, they found such a machine already on the table. The time traveller puts his machine close beside it, sends it back to Thursday, at which point it vanishes: when his friend asks, 'Where's it gone?', he points to the one that has been there all the time, and says, 'That is it'. In this case, the cleaning lady put it back on the table after polishing, but not exactly in the same place. Whatever the details, the time machine was bilocated during the period from 6 p.m. on Thursday to 6 p.m. on Friday.

The other possibility is that in which no-one came into the room during that period. In this case the machine will already be on the table, when the two friends enter the room, and in the exact position from which the time traveller is to set it in action. He must produce his exemplar of the machine from his attaché case, and, having set the controls, hold it over the other exemplar and, at precisely the right moment, press the 'On' button and let it fall. To the friend's eye, it will merge with the machine already there. This is puzzling enough; more puzzling is the apparent ability of the time traveller, by choosing to set his machine beside that already present rather than to engage in the delicate operation I have just described, to bring it about that someone or something moved the machine during the preceding 24 hours.

Both with the forward and with the backward jump, we are concerned with a causal sequence: the machine carries a watch in order to make the sequence clear. Without such a device, a gap in the existence of an object is, on the face of it, a phenomenon symmetrical with respect to past and future; but we should always prefer to interpret such a gap as a forward than as a backward jump in time, especially if the object reappeared in the place at which it had vanished. This is because we take the career of the object to run in the direction earlier-to-later: its position at the moment of disappearance is explained by its career up to that point, not by anything that happens subsequently, whereas its position at the moment of reappearance requires explanation. Likewise, we shall prefer to interpret a period of bilocation, introduced by the sudden appearance of the object at a new place, and ended by a coalescence of the two exemplars at that place, as a backward jump in time.

Backward movement that traverses the intervening time is even more difficult to envisage. In this case, not only must the physiological processes of the time traveller be slowed down, from the standpoint of the ordinary observer, but they must be reversed: his blood, for instance, must collect oxygen from all over his body and deliver it to his lungs, from where he expels it through his nose or mouth. A visitor from the future — say from 2085 — will make an abrupt appearance in his time machine; more exactly, two exemplars of the machine must appear side by side simultaneously. From one of them the visitor steps out, leaving within the other a similacrum of himself whose bodily processes are slowed down and reversed in the way described: provided that no-one tampers with its controls, the machine from which he stepped henceforward behaves in an unsurprising way, and, if carefully preserved, may survive intact (say in a museum) until 2085. If we suppose that the visitor does not return to the time from which he came, he (the exemplar not in the machine) will die in, say, 2020, having been born in, say, 2045. From the latter date onwards, therefore, there will again be two exemplars of him; the younger of them constructs the time machine and positions it. Note that he cannot leave it standing for any length of time and then climb in and switch it on, since it would then have been in the way of itself. He has, rather, to set its controls, have it hoisted above the correct position — which is occupied by that one of its other two exemplars which is not in the museum, but contains himself living backwards — climb in, give the signal for it to be dropped and simultaneously press the 'On' button: he and it will then coalesce with their other exemplars and vanish, leaving only the surviving machine in the museum.

II

I now turn to a philosophical argument which you are unlikely to hear propounded by any philosopher. It is, rather, a piece of popular philosophy. This is the fatalist argument to show it to be pointless to attempt to bring about, or to prevent, any future event. One version of this argument relies on the assumption of determinism — the assumption, in conflict with modern physics, that every detail of what will occur at any future time is causally determined by the present state of the universe. I wish, however, to consider, not this, but the crudest version of the fatalist argument, which relies, not on causal predetermination, but simply on truth and falsity. This was frequently applied, during the bombing of London in the Second World War, to being killed by a bomb. So applied, it ran like this. Either your name is written on any given bomb, or it is not. If it is, then you will be killed by that bomb whatever precautions you take. If it is not, you will survive the explosion, whatever precautions you neglect. In the first case, any precautions you take will be fruitless; in the second, they will be redundant: hence it is pointless to take precautions

Proponents of this argument were not superstitious enough to believe that anyone's name was literally written on a bomb: they employed this figure of speech to mean, 'The bomb is going to kill you'. The general form of the argument is therefore this. Concerning any future event, either it will take place or it will not. If it is going to take place, any action taken to bring it about will be redundant and any taken to prevent it will be fruitless. If it is not going to take place, any action taken to bring it about will be fruitless and any taken to prevent it redundant. Hence it is always pointless to attempt either to bring about or to prevent any future event.

This fatalist argument is not to be dignified by the name of 'paradox'. A paradox, in the strict sense, is an apparently cogent argument, based on apparently true premises, leading to a palpably false conclusion, such as that Achilles cannot overtake the tortoise. In a wider use, the term is applied also to valid arguments, based on true premises, leading to apparently false, or highly surprising, conclusions: it is in this use that people speak of the paradox of voting. But the fatalist argument is a blatant piece of sophistry: no-one would accept this reasoning unless he had a strong motive to believe the conclusion. There is, nevertheless, a good reason not just to dismiss it without further attention, but to try to locate the fallacy exactly.

One attempt would be to deny that statements about the future — statements to the effect that a given event will or will not occur within a given interval — are either true or false. Someone who takes this line may, after a moment's reflection, be willing to grant that they are either-true-or-false, but still deny that they are either: true, or: false. For example, you will either be killed or not be killed by the bomb, but you won't *either* be killed by it *or* survive it. This sounds incoherent. It is an attempt to express an understanding of the connective 'or' according to which a disjunction (an 'A or B' statement) may be true even though neither of the two subordinate sentences is true. A well-known logician, W. V. Quine, once wrote of this as a 'fantasy': but in fact there is no difficulty, for various applications, in constructing a plausible logic which allows it to occur. (The logic of vague statements provides a possible example, quantum logic another.) On the other hand, the idea does not, in my view, yield an effective rebuttal of the fatalist, who will probably win the dispute if his opponent adopts this line. But the principal objection to this strategy is that it is far too heavy for the task in view. It does not expose a fallacy recognizable by all, but involves a questionable revision of the logical principles we ordinarily accept; and it does so on the basis of the metaphysical idea that the future is, as it were, unavailable to render statements about it true or false. Surely the fatalist's argument was a sophistry uncoverable without wading into the deep waters of metaphysics or the foundations of logic.

A natural alternative attempt consists in saying that it may be precisely the precautions that I take that are responsible for my escaping the bomb: the fatalist cannot therefore be justified in inferring, from the hypothesis that I shall not be killed, that I shall not be killed even if I take no precautions, since otherwise he would be entitled to draw the same conclusion from the assumption that I shall escape death thanks to the precautions I am going to take, and to them alone. Suppose, for instance, that, when I emerge with the rest from the shelter, we find that everyone within a half-mile radius of my house has been killed. I shall then very naturally say, 'If I hadn't taken shelter, I should have been killed'; and I mean this to imply that, if anyone had said of me beforehand, 'If he does not take shelter, he will be killed', he would have spoken truly. It cannot therefore have been true to say, as the fatalist would have us suppose, that if I had not taken shelter, I should *not* have been killed.

This is much more along the right lines, but it has two defects. First, if accepted, it shows that the fatalist argument must be wrong, which we probably already believed it to be; but it fails to show exactly where it went wrong. Secondly, it leads us into another morass: quite a shallow one, compared to metaphysics and the validity of the law of excluded middle, but a morass all the same. This is the interpretation of 'if . . . then . . .' statements: the objector will soon get bogged down in arguments about what they mean and in what circumstances they are true. We shall do better not to call in question the fatalist's principle, according to which the truth of a conditional follows from the assumed truth of its consequent. 'If . . . then . . .' certainly can be used in that way, however else it can be used; the fatalist has therefore not yet made a mistake by so using it.

Where, then, do we stand? Let us call the event I want to bring about, surviving the raid in our example, 'E', and the action I propose to take, in our case taking refuge in the shelter, 'A'. The fatalist begins by saying:

Either E will occur or it will not.

Whatever we think about it, we have agreed not to call this in question. He goes on to say:

If E does not occur, it will not occur even if you do A.
If E does occur, it will occur even if you omit to do A.

We have agreed not to object to these statements, either, that is, to interpret 'if', as it occurs in them, so that they are true. The fatalist proceeds:

If E does not occur, A, if you did it, will have been fruitless.
If E does occur, A, if you did it, will have been redundant.

The first of these two statements can hardly be challenged: if I took shelter, but was nevertheless killed, my taking shelter was fruitless. In general, an action done for a certain purpose, and for that alone, must be fruitless when the purpose is unfulfilled: that is what it *is* for an action to be fruitless. It is otherwise with the fatalist's second claim: an action performed for a purpose that is in fact realized is not thereby redundant. A is redundant, when E occurs, only if E would have occurred even if A had not been performed: this is what the second objector was aiming at. The fatalist has not shown this to be so: he has claimed only that, if E is going to occur, it will occur even if you do not do A. This indicative conditional, being of a kind whose truth is guaranteed by the truth of its consequent, will not support the corresponding subjunctive conditional. The fatalist has in fact made no claim that can be embodied in a subjunctive conditional; and so he has no right to assert that, when E occurs, A, if performed, will be redundant.

This, now, seems a satisfactory refutation of the fatalist argument; and you may therefore be breathing a sigh of relief that we must be going to move on from this tedious sophistry. I am, however, going to try your patience a little longer: I should rather a cleaner refutation. The one just given is messy because it appeals to a subjunctive conditional. Subjunctive conditionals crop up all over the place: all sorts of notions, when we attempt to analyse them, appear to depend upon them. And yet the meaning of a subjunctive conditional is enormously obscure; they have been a thorn in the side of analytic philosophers for many decades. Even more obscure than their meaning is their point: when we have devised some theory to explain their truth conditions, we are at a loss to explain for what purpose we want to have sentences with those truth conditions in the language at all. Very important things may hang on them: whether the accused is guilty of murder may depend on whether the deceased would still have died if the accused had not acted as he did. But that is no answer to the question about their point: for, in such a case, we have *given* the subjunctive conditional such a point, and the question is why we should do so.

I should therefore like to be able to state the refutation of the fatalist argument without invoking so opaque though everyday a form of expression. To dwell on this a little, let me recall to you a paradox with which some readers may be familiar. In problem cases, when we are faced with conflicting considerations, it is hard to make up our minds which counterfactual conditionals we should count as true. Moreover, even when we have made up our minds, it is difficult to be sure that the basis on which we decide their truth and falsity is the very same as the basis on which actions are to be judged rational or irrational. Rationality is what the fatalist is talking about: whether it is rational to take precautions or to try to bring something about or prevent it. Now that is very well brought out, it seems to me, by Newcomb's paradox, which I shall here present in a slightly modified version.

Imagine that a rich and eccentric psychologist offers an opportunity for acquiring money to selected individuals. In each case he first interviews the candidate, and then informs him that he has placed equal sums of money in each of two boxes. The candidate is free to choose whether to open only one box or both of them; in either case, he keeps the contents of any box that he has opened. He must, however, make his choice in advance; he cannot first open one box and then decide whether or not to open the other. If the psychologist estimates, on the basis of the interview, that the candidate will choose to open both boxes, he puts £10 in each box; if he estimates that he will choose to open only one, he puts £1000 in each. Over many trials, the psychologist has never lost more than £1000 to any one candidate, nor less than £20: he has got it right every time, even with candidates who said in the interview that they would choose to open one, or both, boxes and then, in the event, made the opposite choice. All this is explained to the candidate.

Smith, when his turn comes, argues, 'The money is already in the boxes. Whether each contains £10 or £1000, I shall get twice as much if I choose to open both boxes'; he accordingly does so, and receives £20 in all. Jones says to him, 'You were a fool: you should have chosen to open only one box', but Smith replies, 'If I had opened only one, I should have got no more than £10'. Jones may well retort, 'Not at all: if you had opened only one, the psychologist would previously have put £1000 in each'. Now which is right? If the question is, which of their judgements about the truth of the counterfactual conditionals best accords with our established ways of judging such matters, it is arguable that Smith has the better case. If so, all that follows is that he has a more accurate grasp of the existing use of such sentences: it does not show that Jones will be acting foolishly when his turn comes, and he chooses to open only one box, obtaining £1000. He thinks to himself, 'I am free to open both or only one. If I choose to open only one, I shall reduce to zero the probability that I shall receive £2000, but shall vastly increase the probability that I shall receive at least £1000, and vastly reduce the probability that I shall receive no more than £20. Since it is in my power to affect the probabilities in this way, it would be irrational of me not to do so'. His reasoning does not turn on the truth or otherwise of subjunctive conditionals, but on factors that it is proper to invoke in determining what he has a rational motive to do; and he ends up with £1000 to prove it. (See the Appendix for a defence of this line of reasoning against an attempted rebuttal by John Mackie.)

Let us, then, return to the fatalist and try to avoid using these subjunctive conditionals in refuting him. One thing is evident concerning subjunctive conditionals, namely that, though distinct in meaning, they are often closely related to the corresponding indicative ones. A subjunctive conditional is often to be counted as true just in case, in different circumstances, an assertion of the corresponding indicative conditional would have been warranted. Thus we have already remarked that, if I later truly say, 'If I hadn't taken shelter, I should have been killed', this implies that, if I had earlier said, 'If I do not take shelter, I shall be killed', I should have been right. Some kind of converse holds also, but to inquire exactly what would land us back in the morass of trying to explain subjunctive conditionals, which we are anxious to avoid. The fatalist is not, after all, interested in subjunctive conditionals, but is trying to persuade us that it is never reasonable to do anything in order to bring about, or prevent, a subsequent event. He is therefore not really concerned with whether or not we say later that the action was redundant, which indeed turns on the truth of a subjunctive conditional: he is arguing that we can never have a reasonable motive for doing it at the time.

So when do we consider that we have a reasonable motive for doing

A in order that E should later occur? Given that we want E to occur, and that it is reasonable to want that, one condition will be satisfied if we have good ground to take doing A as a sufficient and non-superfluous condition of E's occurring. We shall have good ground to take it as a sufficient condition if we have good reason to assert that, if we do A, E will occur; and we may take it as a non-superfluous condition if we have no ground for asserting that, if we do not do A, E will occur — still more, if we have a ground for asserting that, if we do not do A, E will not occur. If we have such grounds, all the fatalist can pit against them is the observation that, if E is going to occur, then it will be true to say that, even if we do not do A, E will occur. Now the supposition that the indicative conditional, 'If we do not do A, E will occur', is true is no obstacle to the truth of the conditional, 'If we do A, E will occur'; for, since these are conditionals whose truth is guaranteed by the truth of their consequents, both can be true together. The fatalist is therefore not, at least on this horn of his dilemma, attempting to undermine our treating A as a sufficient condition of E. He wishes only to undermine our treating it as a non-superfluous condition. He intends us to reason thus: If we knew that the conditional, 'If we do not do A, E will occur', was true, we should regard A as superfluous. But whether A is superfluous or not cannot depend on whether we *know* the conditional to be true: that only determines whether we *think* it to be superfluous. Hence the supposition that the conditional is true implies that A *is* superfluous. Of course, we cannot conclude outright that it is, since we do not know whether the conditional is true or not. We do, however, know that either that conditional or one whose truth implies that it will be fruitless to do A must be true: and hence we know that doing A must either be fruitless or superfluous.

That is how he wants us to argue: but we should, of course, resist such a train of thought. Unlike the similar notion of being redundant we considered earlier, no explanation of what it is for an action to *be* superfluous is available. It is not available since it could not be provided without the use of subjunctive conditionals, which we are eschewing: whether, when we explain such a notion by appeal to them, we have really supplied a firm foundation for it is yet another question I shall not attempt to answer. In the present setting, we have no use for the notion of an action's *being* superfluous, but only for that of our reasonably regarding it as superfluous. The difference is that whether we can reasonably so regard it or not is a matter of what we know or have grounds for believing: no argument can therefore be put forward to the effect that it must in fact be superfluous if a certain conditional, which we have no grounds for taking to be true, in fact holds good.

To have a reasonable motive for doing A in order that E should later occur, we do not, of course, need a ground for regarding doing A as a sufficient condition for the occurrence of E: it is quite enough that we have grounds for regarding it as more probable that E will occur if we do A than if we do not. That, however, is not the only condition: we must also regard ourselves as free to do A or not as we choose. Again, that is overstated: we need only reasonably suppose that we are likely to succeed if we try. More exactly, we must have no ground for thinking that we shall fail save for some reason independent of whether E occurs or not. Normally, I regard starting my car as something I am free to do if I choose; but if, one morning, I find the battery flat, I might ask my son to have it charged. When I return in the evening, I may try to start it. Succeeding in doing so is very nearly a sufficient condition of the battery's having been charged; but I do not regard myself as starting the car in order that the battery should have been charged, but as seeing whether I can start it in order to find out whether the battery was charged. That is because, in this case, I do not suppose that I am free to start the car if I choose, or that, if I fail to do so, this will be explicable independently of the event in which I am interested.

It is natural to think that the topic I am aiming to discuss raises the issue of free will. That issue I believe to be a very difficult and important one, about which most that has been written, from any standpoint, is gravely inadequate; and I do not wish either to deny or to assert that, in any fully adequate discussion of the present topic, it would have to be considered. I believe, however, that it is possible to get a very long way with the surrogate notion of someone's reasonably regarding himself as free to do something if he chooses. Whether he reasonably so regards himself is not a question of whether he has free will: it is a purely empirical question, turning principally on whether or not he has experienced difficulties in performing the action in the past, and on the character of those difficulties, and, secondarily, on the experience of others in this regard.

III

I shall say nothing further about the fallacies of fatalism; but I hope that, despite its being quite obvious that it is fallacious, you have managed to attend to what I have said to show it oo, oinco, as I said at the outset, there is a good reason for exploring this matter. This reason I have so far left unstated. It is this. When people — including philosophers — consider the question whether it would ever be possible, or even reasonable to try, to affect the past, they frequently come out with an argument that is the exact analogue of the fatalist argument. That is to say, their doing this is the best you can usually hope for. If you are not so lucky, they will say things like, 'You can't change the past: if something has happened, it has happened'. But, if you are lucky, they will see straight away that the question is not one of changing the past in the sense of making something that yesterday had happened last week become, today, something that did not happen in that week, any more than affecting the future consists in making something which yesterday was not going to happen next week become, today, something that will happen next week. It is those who avoid this confusion who put forward the analogue of the fatalist argument.

How do they argue? Suppose it is proposed to perform an action B in order that a previous event F should have occurred. Those I have in mind open their argument by observing that either F has occurred or it has not. If it has occurred, they say, it has occurred whether we do B or not: in this case, then, doing B is redundant. If F has not occurred, then doing B, for this purpose, will be fruitless. In either case, therefore, it is pointless: we can do nothing, and cannot ever reasonably suppose that we can do anything, to bring it about that something has happened.

I said that the fatalist argument is not a paradox: it is too obviously fallacious for that. But its analogue, used to show that we cannot affect the past, might be called an anti-paradox. The conclusion seems as obviously true as that of a paradox seems obviously false; but the argument appears valid only for this reason. The argument, being the exact analogue of the fatalist's, is as grossly fallacious as his. It nevertheless impresses a large number of people, including professional philosophers, who, ostensibly on the strength of it, thereupon dismiss the entire issue as unworthy of further consideration.

It thus seems to follow that, in certain circumstances, if we were lucky enough to discover them, it would be reasonable for us to perform an action B in order that a previous event F should have occurred. These circumstances are as follows:

1   The performance of B approximates, in our experience, to being a sufficient condition for the previous occurrence of F: more exactly, there is a sufficiently high positive correlation between them for the performance of B, in a case in which we do not otherwise know whether F has occurred, significantly to increase the probability that it has.

2   We can find no ordinary causal explanation for the occurrence of F, on those occasions when it is followed by the performance of B; or, at least, we can discover no ordinary causal connection between F and B which would explain why B should be correlated with F.

3   So far as our experience goes, B is an action which it is in our power to perform if we choose. That is to say, we do not have, or only very rarely have, the experience of trying to perform B and failing; and, when we do have it, an explanation for our failure is readily forthcoming that makes no appeal to the previous non-occurrence of F.

Now it seems that these conditions would at most provide a motive for the action B in circumstances in which it was not known whether F had occurred or not: for, if it were known, the analogue of the fatalist argument would apply. If F were known to have occurred, it would appear superfluous to perform B; if it were known not to have occurred, it would appear fruitless. But this suggests that it could easily be arranged that one of the conditions 1–3 was violated, provided that it was possible to perform the action B in cases in which it was known that F had not occurred. If it were not possible, this might be regarded as a violation of condition 3; but suppose it possible. The challenge may now be issued to perform B when F is known not to have occurred. Nothing analogous can happen in the case of an action A designed to bring about a subsequent event E, since our knowledge of the future is in part based on our intentions. If we take ourselves to know, independently of our intentions, including the intention to perform A or not to perform it, that E will not happen, that can only be because we do not suppose that the performance of A can, in the circumstances, bring about the occurrence of E; if, on the other hand, our conviction that E will not take place is based on our intention not to do A, we cannot accept a challenge to do A without losing our ground for supposing that E will not occur.

If we accept the challenge to perform B when F is known not to have occurred, it seems that one of two things must happen: either we shall do B, in which case the correlation of B with F will be weakened; or we shall find that we cannot do B, in which case our belief that it is an action which it is in our power to do is undermined. Now, of course, a single trial may not have a significant effect. A single instance in which a performance of B was not preceded by F will not greatly lower the increase in the probability of F which we attribute to a performance of B when it is not known whether F took place or not; and a failure to perform B owing to causes apparently quite unconnected with the non-occurrence of F may be dismissed as coincidence. The failure may not even have the appearance of an unsuccessful attempt: an urgent summons to a forgotten appointment may account for our not even making the attempt.

The situation changes significantly only if there are repeated trials. We may assume that the outcomes are mixed: sometimes B is performed; on other occasions something of an explicable and apparently unconnected kind occurs to prevent the attempt or frustrate its success. Conditions 1 and 3 will not be sufficiently preserved, however, to justify maintenance of the belief that doing B is a way of bringing it about that F occurred if all the outcomes are of one or other of these kinds: we must suppose that a third type of outcome occurs with significant frequency. This is that B is performed, but it subsequently proves that the information that F had not happened was erroneous, owing to an error of observation or in the transmission of the report. If this happens often enough to enable conditions 1−3 to be preserved to a degree sufficiently great to warrant maintenance of the belief in the efficacy of B, it will tend to change our attitude to evidence that F did or did not occur. Originally we made the natural assumption that we could, on occasion, know whether or not F had occurred independently of our intentions — precisely the assumption that cannot be made in relation to any future event which we believe ourselves to have the means of bringing about or of preventing. It was because we made that assumption that we were able to establish the correlation between the performance of B and the previous occurrence of F. It was also because we made this assumption that we took it for granted that there was no point in trying to bring it about that F occurred when we have clear evidence that it did not. Although we placed sufficient reliance on the correlation between B and F for the performance of B to count as increasing the probability that F occurred in cases in which we had no evidence of an ordinary kind about whether it did or not, we trusted such evidence so much more than we trusted the correlation that the performance of B did not significantly affect our estimate of the probability of F's having occurred in cases in which we possessed that evidence, even though we knew that evidence for a past event can sometimes prove mistaken. Now, however, we shall be impelled to abandon this assumption. The performance of B is now felt to increase the probability that any report we have that F did not occur will be found to be in error: we therefore no longer believe that the performance of B is irrelevant to our judgement of the probability of F's having occurred when we have other grounds to go on. Since we still regard performing B as being, by and large, within our power to do when we choose, it follows that the formation of an intention to do B is likewise relevant to such a judgement. In the intermediate stage, the performance of B merely increased the likelihood that a report that F had not occurred would be found to have been in error: but the final result will be that we shall take that action, or even the formation of an intention to perform it, as itself grounds for doubting any such report, even in the absence of any contrary evidence of the usual kind.

Thus what stands in the way of our supposing it rational to do anything in order that something else should previously have occurred is not the logical fact that the event in question has already either occurred or not occurred, or the metaphysical status of the past as fixed, in contrast to the fluid condition of the future, but our assumption that, of any past event, we may have evidence for its occurrence or non-occurrence whose strength can be estimated independently of our intentions. This assumption is, of course, based, not only on causal connections from earlier to later, but on the absence of any comparable connections in the reverse direction — that is, connections that we might use to attempt to bring it about that certain events had previously occurred. Just because the assumption is so deeply engrained in us, we should feel the strongest psychological resistance to recognizing any such connection; but, were we to recognize one, we should have, to that extent, to modify that assumption. This would profoundly alter our conception of evidence about the past, but it would not produce conceptual chaos.

IV

What is our conception of the temporal priority of cause over effect?

Not, plainly, that there must be a temporal gap between an immediate cause and its effect, but, rather, that a causal chain runs always in the earlier to later direction. Each link in the chain is a process, whose initiation is the immediate, and thus simultaneous, effect of the arrival at a particular stage of the process that constitutes the preceding link. The subsequent continuation of the process is not treated as requiring causal explanation, but only its deviation, if any, from some norm, which is to be explained by the continuous (and simultaneous) operation of some causal influence. It is the fact that it is the *subsequent* continuation of the process, once initiated, that calls for no explanation which gives a temporal direction to the causal chain.

Can we imagine a world in which the temporal direction of *all* causal processes is reversed? If memory is treated as a natural process, we should lack the faculty of memory in such a world, and have, instead, the analogous faculty of precognition; and it now becomes difficult to say whether, in imagining this, we have not simply switched the words 'past' and 'future', that is, whether we have genuinely imagined a different state of affairs. If, on the other hand, we exempt memory from the time-reversal, then, again, there seems little difficulty if we suppose that we are mere *observers* — a kind of intelligent tree, able to communicate with one another, but to engage in no other action or form of interference with the course of events that we witness. If, on the other hand, we imagine ourselves as interacting with such a world, we shall have to exempt intention, as well as memory, from the reversal of causal direction, if we are to preserve any ordinary conception of action at all, since it is integral to the concept of intention that there be a high correlation between forming intentions and carrying them out. For us to be able to form future intentions at all, we should have to have a cognitive attitude to the future not wholly analogous to our present attitude to the past; it would nevertheless be somewhat analogous to that, in that our knowledge of what was to happen would rest in part on our intentions and in part on inferences from present traces of future events parallel to the inferences we now make concerning past events of which we find present traces. In consequence, we should regard far fewer actions as in our power to perform than we now do. This would mitigate the weakening of our confidence in our memories that would result from the phenomenon on which I have commented, that we should no longer believe, in general, that we could have knowledge of what had happened independent of our intentions. All this makes the thought experiment difficult: but, as far as I can see, it is not impossible.

In such a world, remote causes would be connected to their preceding remote effects by causal chains running in the later-to-earlier direction. In discussing the coherence of the conception of bringing about the past — doing something in order that some subsequent event should have occurred — I ignored the question whether such a chain could be discovered between the action and the earlier event. In doing so, I failed to provide a true analogue to earlier-to-later causal connections. A correlation between temporally separated events can of course be observed, and, so far, provides evidence only for the hypothesis that there is some causal connection between them. That the earlier lies on a direct causal chain linking them can only be fully confirmed, or disconfirmed, by the identification of such a chain, and hence the integration of the connection into our system of causal explanations; but the hypothesis that it does can be supported by observational evidence that the later event is not equally well correlated with an immediately antecedent condition, in cases when the earlier event is inhibited. The same applies when we reverse the temporal order of the two events. It could occur that every attempt to correlate the earlier event with some earlier link in a recognized causal chain leading to the later one breaks down. Indeed, this must happen if the conditions I stated for it to be rational to attempt to bring about the past are to be satisfied. In that case, the later event was a voluntary action: discovering that the earlier event was equally well correlated with some event that was earlier still and led to the performance of the action would destroy the belief that the action was in our power to perform, and turn trying to perform it into a way of finding out whether the earlier event had occurred. But, even so, we have, at best, an analogue of the case in which there is some supporting evidence for the hypothesis that a certain event is a remote cause of some subsequent one, not of that in which that is as secure a belief as any concerning causality can be. For that, we should need a chain linking the later event to the earlier one, composed of processes whose *previous* stages were taken to require no explanation, given what ended them. Is it a coherent idea that we might recognize the existence of such a chain, even though our general system of causal explanations ran, as now, in the earlier-to-later direction?

It is certainly the complete absence of anything to fulfil the role of

such a causal chain that makes parapsychology — at least, that part
of it that treats of alleged instances of precognition — appear utterly
unscientific; but is this in the nature of any purported case of
backwards causation, or due only to the feebleness of parapsychology? The difficulty of the idea plainly decreases the further
we move from the sphere of processes which we are capable of
interrupting. Any attempt to consider anything as simultaneously
subject to influences in both temporal directions is liable to collapse
in complete incoherence. Let us consider the model time machine
once more, and suppose that it travels backwards continuously in
time. Let us first suppose the room to be empty from 5 p.m. on
Thursday until 5.55 p.m. on Friday, when the time traveller and his
friend enter. The camera records the following scene. At 5 p.m. on
Thursday the table is bare. Precisely at 6 two model time machines
appear abruptly on the table side by side; inside one is a watch which
reads 7 p.m. Friday and is running normally, and inside the other one
that gives the same reading and is running backwards at the rate of
two and a half minutes per hour. Both machines continue in the
positions at which they appeared until 6 p.m. on Friday, at which
moment the time traveller, having taken a simulacrum out of a box,
sets its controls and holds it over the machine containing the watch
now telling the right time (the other reads 7 p.m. Saturday), drops it,
pressing the 'On' button: both it and the machine on to which he
drops it vanish, leaving only the machine with the fast watch.

Now suppose that, at 10 o'clock on Friday morning, the cleaning
lady comes in. Is it possible that she should take both exemplars of
the time machine from the table and put them on the mantlepiece?
There is no difficulty about her doing this to the exemplar in which
the watch is running normally: but it is impossible that she should
do it to the one that is travelling backwards in time, or, otherwise
expressed, within which all processes are slowed down and reversed.
She cannot do it, because we have assumed

1   that the time traveller started the machine on its backwards
    journey on the table; and
2   that that journey, which occupies it from 6 p.m. on Friday to
    6 p.m. on Thursday, is a process which has the temporal analogue
    of the property of requiring no explanation for its continuance,
    once initiated.

What is this analogous property? Described in our ordinary earlier-
to-later terms, it is that of being a process which, from a certain

moment t on, will continue until the terminating event. The terminating event in our case is the time traveller's pressing the 'On' button. The moment t will be either the initial event — in our case the sudden appearance of the time machine at 6 p.m. on Thursday — or the very first moment of any intervening action on the machine. It follows that no such intervening action, with *subsequent* causal consequences, such as the cleaning lady's removal of the machine to the mantlepiece, can occur: for those consequences cannot ensue, given the stated assumption about the character of the process the machine is undergoing during those 24 hours.

We must not conclude that, while it is undergoing this process, the machine has become an immovable object: the impossibility is a logical, not a physical, one. What made the supposition that the cleaning lady moved the machine impossible was our assumption that, when the time traveller and his friend entered the room, the machine was on the table, and that he pressed the 'On' button as he let it fall to the table. The following sequence is consistent. At 6 p.m. on Thursday, the machine (or, rather, the two exemplars of it) appears suddenly on the table. The cleaning lady comes in the next morning, and removes the machine undergoing backwards travel to the mantlepiece, and the other one to a shelf. That evening, the two friends find the two exemplars in those positions, and, as he sets the 'On' button, the time traveller drops his exemplar on to the one on the mantlepiece.

This last is, I think, a coherent description; but it has puzzling consequences. Why was the machine on the mantlepiece when the time traveller and his friend entered the room? Its position at that time is doubly determined: by the fact that the cleaning lady put it there; and by the fact that the time traveller subsequently set it going there. On the other hand, there is no reason at all for its being on the table at the moment when the cleaning lady picked it up. Its presence on the table is not accounted for by the cleaning lady's picking it up, which serves only to explain its *subsequent* position; and it is not accounted for by its having previously been on the table, since it was at that time undergoing its backward journey, any phase of which serves only to explain *previous* phases. The price of admitting things to be subject to causal influence in both temporal directions may be that for some phenomena we have a multiple explanation, while for others we have none.

V

If there are causal chains running in the reverse as well as in the usual direction, there is a possibility of causal loops. They often occur in science fiction: one such story that amused me concerned a fifth-rate but conceited artist. One day he is visited by an art critic from a century ahead, who explains that he has been selected for time travel so that he could interview the artist, who is regarded, in the critic's time, as by far the greatest artist of the twentieth century. When the artist proudly produces his paintings for inspection, the critic's face falls, and he says, in an embarrassed manner, that the artist cannot yet have struck the inspired vein in which he painted his (subsequently) celebrated masterpieces, and produces a portfolio of reproductions that he has brought with him. The critic has to leave, being permitted, for some unstated reason, only to remain for a limited length of time in the past, and the artist manages to conceal the portfolio, so that the critic has to leave without it. The artist then spends the rest of his life producing the originals of the reproductions by carefully copying them in paint.

The natural reaction to such a cycle is that there is no explanation for *its* occurrence. The existence of the reproductions is to be explained, in the usual way, by reference to the existence of the originals; and the existence of the originals can likewise be explained by reference to that of the reproductions from which they were copied: but there is no reason whatever for their joint existence — no reason why there should be any paintings and reproductions like that.

Reflection on such examples throws a little light upon a puzzling feature of the philosophy of St Thomas Aquinas. In his proof for the existence of God as First Cause, Thomas says that you cannot go back to infinity in a sequence of causes, that is, that you cannot have an infinite descending causal chain, where by a descending causal chain I mean a sequence each term of which other than the first is a cause of the preceding term. The thought is evidently modelled on the principle that you cannot have an infinite descending chain of (cogent) grounds. A proof, if it is to carry conviction, must have premisses. It might be possible to give a precise characterization of an infinite deductive structure, in the form of a tree: each statement stands below one or more other statements from which it follows by a

valid principle of inference. Since it is infinite, however, there must
be at least one branch of the tree that does not terminate, that is, has
no tip; and, if so, the proof establishes nothing. The point of a
deductive proof is that truth is transmitted downwards: given the
truth of the premises, the truth of the conclusion is guaranteed. In
the infinite proof, we have no reason for accepting any of the
statements occurring along the infinite branch as true, and hence no
reason whatever for accepting the conclusion. It is natural to think
that Thomas regarded causes in the same light: an infinite descending
causal chain would offer *no* explanation for the existence or occurrence
of the first term in the chain.

What is puzzling is that Thomas elsewhere argues that we have no
ground in reason for believing the universe to be of finite age (he
thought that revelation supplied such a ground). Why, then, can we
not prove that the universe must be of finite age by appealing to the
impossibility of an infinite descending causal chain? And, if we
cannot, why can we not refuse the conclusion to the existence of God
as First Cause by proposing that the universe has always existed?
The standard response of the commentators to the latter question is
to say that we require an explanation of the existence of the sequence
as a whole. I used to think this response absurd in itself, and lacking
any warrant in the texts. Whether warranted or not, it cannot be
dismissed offhand as absurd in itself. A causal cycle of length n is a
special case of an infinite descending causal chain, one in which every
block of n terms repeats; and the intuitive reaction to the supposition
that there could be such a cycle is precisely that it would leave the
existence of the cycle as a whole unexplained.

There is a passage in Thomas's writings in which he attempts to
resolve the problem caused by his denial of the absurdity of a
universe without temporal beginning by distinguishing two types of
descending causal chain, one of which may be infinite and the other
of which cannot. In making the distinction, he employs the terms *per
se* and *per accidens*, which seem capable of a wide variety of
applications in medieval philosophy. The idea is that it is intrinsic to
the existence of certain things that they should have been produced
in a certain way, whereas in other cases it is merely a contingent fact
that they are always produced in a particular manner: when the
chain is of the former type, it must be finite, but when of the latter, it
may be infinite. As a specific example of the latter type, Thomas
cites a chain composed of human beings, each term of which other
than the first is a parent of the preceding term. It is not intrinsic to

the existence of human beings that they have parents, Thomas thinks: after all, Adam and Eve had none. The difficulty is to cite examples of the former type. Here is one that is attributed to Wittgenstein. You come upon a man, and hear him say, 'Four, three, two, one, nought. Whew!!'. You question him, and he replies, 'From eternity I have been running through the sequence of natural numbers backwards, and I have just got to the end'. It is plain that this strikes us as conceptually absurd in a way that its opposite does not, or at least as measurably *more* absurd than its opposite. If there were a Greek legend of a sinful mathematician condemned in Hades to count through the natural numbers for eternity in ascending order, this would appear far more intelligible than Wittgenstein's fantasy. A causal sequence of the order-type of the positive integers appears less baffling than one of the order-type of the negative integers: why? I suppose the thought behind the intuitive reaction is this. When the sinful mathematician says, '31', for example, at a particular moment, his reason is that he had just said '30': and his reason for saying that can be traced back to his reason for saying 'Nought' at the outset, which in turn lay in the sentence of the infernal judge. In the same way, the reason for Wittgenstein's friend having said '31' at a particular moment lay in his having just previously said '32'. *This* reason, however, cannot be traced back to anything, and so he can have no reason to be engaged in the countdown at all: he has no reason to utter the name of any particular number because there is no reason why he should be engaged in the entire process. Once more, it seems that there can be no adequate ground for the occurrence of any term in the sequence, because there is no ground for the occurrence of the sequence as a whole.

VI

I am not going to attempt to resolve this perplexing matter. I shall not end, as Wittgenstein once ended a lecture, by saying, 'For the moment, I shall leave you puzzled': there is no 'for the moment' about it, because I am deeply puzzled myself, and, if I tried to take the matter further, should probably talk utter nonsense. I shall end only by remarking that there is a possible means of affecting the past that I have not considered, namely if time is itself cyclic, so that the recent past is also the remote future and conversely. In this case, of course, the terms 'past' and 'future' do not, relative to any moment and frame of reference, distinguish disjoint temporal regions; hence, even if all causal chains run in the same temporal direction, anything that affects what subsequently happens thereby also affects what has happened. It does not follow from that that there are causal loops, since all effects might dissipate within a small fraction of the length of the entire cycle; but the possibility of loops is open, leaving us with the puzzles I have confessed that I do not know how to resolve.

DISCUSSION

Q. There's a stock objection to backwards causation which has to do with people killing their grandfathers and so forth. You seem to be trying to avoid this problem by saying that, in some sense, the probability of a past event always falls short of certainty, that you can't actually *know* that a past event happened. And that seems a very counterintuitive thing to say.

A. Well actually, I didn't even say that there were not events of which you *could* know for certain that they had happened. But in general it *is* a matter of probabilities and assessing reports that we get. The situation I envisaged was this. There's a type of event F and a type of action B. You believe on the basis of past experience that you're free to perform B or not as you choose. Moreover, you regard B as a sufficient condition of the previous occurrence of F. That's also based on empirical evidence. Finally, you've got no other explanation for the occurrence of F, no ordinary causal explanation that accounts for the correlation that links them. Those were the assumptions. The question then is whether these assumptions are cotenable. Someone who thought they were not might issue the following challenge: you can show that you can't hold all of these together if you keep trying to perform B in conditions in which you take yourself to know that F has not occurred. Now perhaps in some cases responding to that challenge *would* violate one or other of those beliefs. You might find yourself just unable to do B in many of those cases. Then you would give up your belief that you were free to do B or not. Or you might find that you frequently did it. And there it was: you've got B being performed on many occasions when F has not occurred. That weakens the belief in the correlation, perhaps to vanishing point. I'm not denying that those are possibilities.

What I was considering, however, was whether things could so turn out that, even though you respond to this challenge, these conditions conditions in which I maintained that it would be rational to do B in order that F should have occurred — still sufficiently persist. And what I said was that this can only happen if it frequently then turns out that the report that F did not occur was erroneous. Now you can't deny that very often we do find that we've made mistakes about what has happened in the past. One just can't legislate that out of existence. So I'm entitled to suppose that, in a hypothetical case, that happens sufficiently often so as not to destroy, though it may weaken, the other assumptions. What it then does is a very peculiar thing. And that, I think, helps to explain why we have such a resistance to the idea of backwards causation, or doing something in order that something else should have happened. We find that what gets weakened is our normal assumption that you can tell whether or not an event has occurred independently of your intentions. And now the intention to do B becomes itself a ground, in some cases, for supposing that F has occurred, even when there are no other reasons. Of course not from the beginning. It's just that finding so often, when we do B, that it turns out that F has, after all, occurred puts you in that frame of mind.

Now that's a very big conceptual change. But I don't think it produces conceptual chaos. A lot of people think that the idea of backwards causation — at any rate backwards causation that we could get a handle on in order to bring about what has happened — is complete nonsense. But I don't think so. It's certainly not nonsense because of the analogue of the fatalist argument. If the fatalist argument is no good in the hands of fatalists, then it's no good here.

Q. Logicians seem concerned to rescue ordinary people from the pathological consequences of logic choppers who are distorting language. And they do this by sophisticated arguments which are, ultimately, appeals to common sense. Now we've been hearing in other lectures about causality from the point of view of quantum theory, where we get paradoxes like Schrödinger's cat. I wondered how a logician would regard that. Is that kind of causality outside the preserve of logicians?

A. Not at all. Nothing is outside the preserves of logicians. That's by definition. If logic doesn't cover every realm of discourse, then it isn't logic. But continue with your question. (I'm not going to talk about Schrödinger's cat, if that's what you were hoping.)

Q. I wondered if you felt that these paradoxes of quantum theory could also eventually be reduced to commonsense terms, if we had sufficient insight into the common usage of such expressions as 'causality'.

A. I think that's a misconception of what I said. I do not think that it's a matter of common usage. Quite the contrary. Of course you have to respect common usage to some extent. Those words you use and don't specially explain are best used in accordance with their accepted meanings. Otherwise you'll be misunderstood or will make claims that are quite unjustified. But solutions to philosophical problems that turn simply on appeals to common usage are, for the most part, spurious in my opinion. And that is precisely why, for example, I tried to get away from these subjunctive conditionals. You can spend a lot of time discussing in what circumstances the statement 'If I had done so-and-so, such-and-such would have happened' is true. And when you've finished doing that, if you're not careful — unless you tie it up in some way — all you've got is an account of the way we normally use conditionals of that form. And it hasn't answered the question you seriously wanted to answer. Particularly is this so if you concentrate on how the word 'cause' is used, and how it's connected with a temporal direction. Now, if someone wants to know whether it's reasonable for him to do something when his motive is that something should have happened — which some people regard as absolutely ridiculous — it's a cheat to fob him off with explanations in terms of how we use the word 'cause'. That's not the question. Whether it would be called a cause, or is rightly called a cause, may be an interesting question, but it is not the question he was asking. If it's not called a cause, then all right, perhaps we'll call it something else. But the question is: is there any sense in doing this thing? So you've misunderstood me if you thought I was making appeals to common usage. And I wasn't making appeals to commonsense either. In fact I thought I was flying in the face of commonsense to some extent.

Q. Can I put something to you that arose in a previous lecture? Paul Davies mentioned a fascinating thought experiment that John Wheeler has been writing about, in which it looks as though there's backwards causation. What you have is a light source, a screen with two apertures, a lens with an image screen in front of it, and two photon detectors. Now the image screen is made up of slats which can be opened like a venetian blind. If the slats are open, the photons will pass through them, and the lens will deflect each photon into the upper or lower detector, depending on whether it has gone through the lower or the upper aperture. If they are closed, on the other hand, one will get interference fringes on the screen, just as in the classic Young's two slit experiment. And the way this would normally be explained is that, when the slats are closed, each photon exhibits wave-like behaviour and goes, so to speak, through both apertures. If the slats are open, however, then, because the detectors enable one to tell which aperture the photon goes through, the accepted wisdom is that it behaves like a particle, and just goes through one aperture or the other. But Wheeler pointed out that you can delay the decision as to whether to have the slats open or shut until after the photon has passed through the screen containing the apertures. Then it looks as though it's a matter of retroactively causing the photon, as the case may be, to behave like a wave or like a particle. But it struck me, listening to your talk, that actually this kind of backwards causation is totally benign in terms of your account. For suppose that you are the experimenter and you want to know, after the photon has negotiated the apertures but before it reaches the image screen, whether it has gone through just one aperture or, as it were, spread out between the two. It is quite clear that there is, in principle, no way of knowing that independently of knowing your own intentions: specifically whether you intend to open (close) the slats, assuming them to be closed (open), or to leave them as they are. I wonder if you have any comments on this.

A.   Well I think you've already made the comment which I was intending to make! As for benign, I don't mean that it's easy to swallow. What is particularly difficult is to put such phenomena into any coherent system of causal explanations. Still, I do think that what you said is absolutely right. As I understand it, there is, in the kind of case you describe, no possibility of generating causal loops. And this is surely a crucial consideration in its favour.

APPENDIX: MACKIE ON NEWCOMB'S PARADOX

Newcomb's paradox involves a case in which I can choose between two courses of action, knowing that one or other of two situations obtains, but not knowing which: the outcome will depend both on my choice and on the situation in fact obtaining. The principle of dominance lays down that if one of the two courses of action will, in *each* of the two situations, produce the more favourable outcome for me, I must rationally choose that course of action. In the version of Newcomb's paradox discussed by the late J. L. Mackie in 'Newcomb's paradox and the direction of causation' (*Canadian Journal of Philosophy*, Vol. VII, 1977, pp. 213—25), one box is open to view, and contains $1000, and the other is closed, and contains either nothing or $10,000: I have the option of either taking only the closed box or both of them, in both cases keeping the contents. The pay-off matrix is therefore as follows:

	the closed box is not empty	the closed box is empty
I take the closed box only	$10,000	$0
I take both boxes	$11,000	$1000

The strategy of taking both boxes dominates that of taking only the closed box, since in either situation I get $1000 more.

Bar-Hillel and Margalit, in 'Newcomb's paradox revisited' (*British Journal for the Philosophy of Science*, Vol. 23, 1972, pp. 295—303), had argued that the principle of dominance fails to apply when the choice of strategy affects the probability that one or other situation obtains. Mackie denies this, and maintains that the principle fails only when the choice of strategy causally determines, or at least influences, the situation. The immediate argument he gives against Bar-Hillel and Margalit is, however, irrelevant. It is given in terms of an example of theirs concerning the relations between Israel and Egypt: but I shall here transpose his argument to the Newcomb context. So transposed, it is as follows. Suppose that, before I make my choice, I get a glimpse of the inside of the closed box, without being sure that I have seen aright. And suppose that, if the closed box appeared to me to be empty, I am strongly moved to take both boxes, whereas, if it appeared to me to contain $10,000, I am moved to take only it. Then my taking only the closed box makes it more probable that it is not empty: but this provides no rational ground for me to depart from the principle of dominance.

Mackie's argument looks plausible, partly because it is very swift, and partly because it is *not* transposed to the Newcomb context: but it is doubly wrong. First, it violates the conditions of the paradox, which involve that I have no basis for judging whether the closed box is empty or not save by reference to my own choice of strategy. Secondly, and crucially, it affects the probabilities as judged by a third party, but not by me. If *you* know that I have had a glimpse of the closed box, but do not know what I (thought I) saw, and if you know how this glimpse is likely to affect my choice of strategy, then my taking only the closed box makes it more likely, for you, that I saw money in it, and hence that it has money in it. Having had the glimpse, however, my choosing only the closed box does not, for *that* reason, make it more likely *for me* that there is money in it, because it makes it not in the least more likely, for me, that I thought I saw money in it: I already know what I thought I saw. Mackie's argument therefore provides no counter-example to the thesis of Bar-Hillel and Margalit, which was that the principle of dominance fails, as a criterion of rational choice, whenever my choice of strategy affects the probability that *I* (rationally) assign to the obtaining of the two possible alternative situations. I shall here defend that thesis, which I shall label the BHM thesis.

The assumption of the paradox is that, in every instance in which someone chose to take both boxes, the closed box proved to be empty, and, in every other instance, it contained $10,000: all those choosing both boxes got $1000, and all those choosing only the closed box got $10,000. Mackie is of course quite right that I have to ask myself whether I can extrapolate. If I decide that this regularity was pure coincidence, I may rationally follow the principle of dominance: on this hypothesis, my choice of strategy will not weigh with me in estimating the probability that the closed box contains $10,000. This case, too, is irrelevant to the BHM thesis.

Mackie goes further, however, and insists that, if I do not dismiss the regularity as coincidental, I must ask myself *how* it occurs. If I decide that someone's choosing to take only the closed box *brings it about* that it contains $10,000, when otherwise it would have been empty, then, Mackie agrees, it will be rational for me to ignore the principle of dominance and choose to take only the closed box. This will happen, Mackie says, if I conjecture that, by some trickery, the $10,000 are inserted into the box after the subject has announced or otherwise demonstrated that he will adopt the strategy of taking only the closed box. He adds that it will also happen if I suppose that backwards causation is operating. Once more, these cases are not at issue between Mackie and Bar-Hillel and Margalit: they are agreed that the principle of dominance fails in cases such as these.

What if I decide that the regularity is not coincidental, but that my choice of strategy does not bring about the condition of the closed box? How can I suppose this? The agent whose money goes into one or both boxes claims to be a psychologist with a deep understanding of character, who puts $10,000 into the closed box if, and only if, he predicts that the subject will choose to take only that box. Now, Mackie says, I can hold that the regularity is not coincidental, without supposing that my choice of strategy brings about the condition of the closed box, by believing the psychologist's claim. In that case, he claims, I have decided that each person is determined by his character to adopt one strategy or the other, and that it is the psychologist's perfect assessment of this character that determines the situation, namely whether the closed box is or is not empty. But, if I think this, Mackie maintains, I must recognize that my choice of strategy is not free: I therefore do not need to make up my mind what to do. More exactly, he does not actually say that I do not need to make up my mind: what he says is that the question, 'What is it reasonable for me to do?', is idle.

Obviously, this is wrong: for me the question is not idle at all. Here I am, wavering between taking both boxes and taking only the closed one, and trying to decide which it is reasonable for me to do. The thought, 'My choice is not really open: the psychologist already knows which I shall do', may fill me with despair: but it will not help me to decide, and it equally will not dispense me from deciding.

A compatibilist, in philosophical jargon, is one who believes that our possession of free will is compatible with complete causal determinism. Mackie remarks that, in a case like that under consideration, I may be free 'in a sense which compatibilists might find sufficient'. He claims, however, that the compatibilist must interpret the question, 'What is it reasonable for me to do on this occasion?', as meaning, 'What sort of action is in general most advantageous in circumstances of the type to which those now obtaining belong?'. On Mackie's own view, these are two quite different questions, although, in most circumstances, they have the same answer. To the second question, he has no doubt of the answer in the present case: it is in general more advantageous to take only the closed box. But the peculiarity of Newcomb's paradox, he argues, is that is splits the two distinct questions apart. In the particular case under consideration, in which I am supposing that each person's choice is determined in advance by his antecedent character, the first question *has* no answer, according to Mackie: there is nothing it is reasonable for me to do, since I believe that my choice is not free.

Mackie's conclusion thus appears to be that, if Bar-Hillel and Margalit are compatibilists, their thesis is after all correct, but only because it does not mean what it appears to mean: being compatibilists, they are forced to put a deviant construction on the notion of rational choice. If the notion is understood in the standard way, however, namely as a non-compatibilist interprets it, then the BHM thesis is incorrect.

The conclusion seems implausible: it makes the solution to Newcomb's paradox turn on the perplexing issue of free will and determinism, whereas it hardly seems that one has to resolve this issue before deciding what is the rational choice for one of Newcomb's subjects to make. Certainly it is unnecessary to have any opinion about the issue to acknowledge that, in the case in question, I have to make up my mind what to do, and shall want to do so rationally. Mackie acknowledges that a determinist could allow that I was capable of being influenced by rational argument: but he takes this only as a breach in the determination of my choice by my character. Perhaps others can advise me before I make my choice of strategy, and perhaps they may persuade me to do what I should not otherwise have done; but, then, there is, after all, a little room between my character and my choice, even though I am not really free.

The question is not, however, whether some third party, who holds determinist opinions, will allow that I might be influenced by rational argument. It is not even whether Mackie can admit that such a determinist could consistently allow this: it is whether *I*, the subject, holding determinist opinions at least about the choice with which I am faced, can, on Mackie's view, consistently seek or entertain such advice. It is obscure how I can. According to Mackie, I ought to regard the question, 'What am I to do?', as idle. What does it mean to regard that question as idle? Perhaps that I should not view myself as having any choice to make. If so, there will, nevertheless, be such a future event as my selection of a strategy: but, perhaps, on Mackie's view, that should not be described as my making a *real* choice, whatever *that* means. Well, whatever it means precisely to say that I should regard the question what I am to do as idle, Mackie certainly intends it to imply that the question what is the rational thing to do does not arise for me: if it did, he would be guilty of failing to pronounce on which strategy would be rational in such a case.

Now, it might be said, even though I believe my selection of a strategy to be determined by my character, I may still consistently suppose that it may be affected by extraneous factors, such as listening to somebody's advice. I may indeed suppose that: but what motive have I, on Mackie's account, for subjecting it to such factors? The only motive I might have is that they would tend to make my choice more rational: but Mackie's position is that, having once adopted a determinist view of my forthcoming selection of a strategy, I can no longer regard it as rational or irrational. Even if someone comes unasked to offer me advice, how can I listen to it and be swayed by it? The effect of advice on my subsequent actions is not an adventitious one, like that of alcohol or a good night's sleep: it is supposed to operate through my understanding of the words and my estimation of their truth. But any advice I receive will be to the effect that one strategy or another is the more rational: and I am expected by Mackie, if I am rational, to suppose that the categories of rationality and irrationality do not apply to my selection of a strategy.

In any case, we can simply set up the paradox to make the intrusion of advisers impossible: after I commit myself to being a subject of the experiment, I have no opportunity to take anyone else's advice until after I have chosen. All the same, I can still revolve the considerations in favour of each strategy in my own mind: perhaps no adviser would be able to put before me any consideration that had not already occurred to me. I still must ask myself, 'What shall I do?': however convinced I may be that my choice is fore-ordained, I still must make it.

Suppose that I explain the fact that the closed box has, in the past, proved empty in just those instances in which the subject chose to take both boxes by appeal to the psychologist's discernment of character, without supposing him infallible: there have been 2000 previous subjects, and I estimate that he is liable to make a mistake once in every 5000. I might still believe in a strict determination of choice by antecedent character; but I might not believe that, thinking that some actions, though only a very few, are out of character. On this view, my choice is then free, though largely constrained; but it will make no sense to say that the question, what I am to do, is *largely* idle. The question is either idle or it is not: either my choice can be described as rational or irrational, or it cannot. Mackie's view is that, in this case, I must, if I am rational, follow the principle of dominance: the small margin of error on the psychologist's part makes all the difference, because, since I have a degree of freedom, even though so small a degree, the question what it is reasonable for me to do now genuinely arises, and, when it does, it is always rational to follow the principle of dominance when the choice of strategy does not causally determine the situation. It would therefore be irrational for me to take into account the fact, as I consider it to be, that the probability that I am the one subject in 5000 about whom the psychologist is mistaken is only 1/5000. Mackie attempts to soften the unconvincing ring of this answer by giving a tendentious description of the case: he says that the subject may be able to 'break free from his own established character by a heroic effort of will', and that if that character tends very strongly to make him take the closed box only, 'he should make a supreme effort to break free from it and take both'. Plainly, however, this need not be a correct account of how the matter appears to the subject at all. The description would fit if he were conscious that, in those circumstances, his judgement was liable to be warped, and was trying to bring to bear a previous resolution, made when it was more reliable. If, on the other hand, he has no particular reason to distrust his present judgement, talk of a supreme effort is inapplicable: if, upon reflection, he is disposed to take only the closed box, he has no motive to act contrary to his judgement.

Still, in considering this last case, Mackie has found himself no longer able to evade the question he has for so long refused to face: and his answer is quite implausible. It is plain that, if the psychologist really is a good predictor of each subject's choice of strategy, those who make the choice Mackie considers to be irrational will do much better than those who make the other choice: and this needs some explaining. He explains it by saying that 'rationality as a character trait, defined as a strong tendency always to do what, if one were free to choose, would then be the more advantageous thing to do, is not necessarily the most advantageous character trait to have'. The definition is faulty: the rational agent will not always do, and may not even tend to do, what is in fact most advantageous, since he may be badly misinformed. Roughly speaking, a wholly rational agent is one who always does what, on the basis of what he knows, is most likely to be the most advantageous thing; more exactly, one who maximizes expected utility. This cannot but be the most advantageous character trait to have, or, more exactly, the most likely to be advantageous. There can be cases in which an imperfectly rational action proves more disastrous than an unintelligent one, as when a chess player's opponent prepares a trap he will fall into only if he is a fairly good player; and there can be cases in which misinformation, or an incorrect assessment of probabilities, will lead a rational agent into worse trouble than an irrational one, because the latter happens to hit on a policy that would have been rational for someone correctly informed: but there cannot be a case in which, on the basis of given information, an irrational policy has a greater chance of being advantageous than a rational one. Any policy compliance with which has a strong probability of not yielding the greatest advantage simply cannot be that which a rational agent will adopt.

The only kind of freedom relevant to the paradox is that I am free to take one box or both in the sense that I can rule out the possibility that I should attempt to take both and find myself unable to, or attempt to take only one and be unable to avoid taking both. Suppose that I believe that I am, in *that* sense, wholly constrained. I think, that is, that, if there is nothing in the closed box, I shall simply be unable, if I try, to take only it, and that, if it contains $10,000, I shall similarly be unable to take both boxes. Then, indeed, the question, 'Which shall I do?', has become idle for me: it will, I believe, turn out the same whichever I decide. But, if I do *not* think that, the question is not idle, independently of any opinions I may hold about determinism. In this case, I regard myself as free, in this, the only relevant, sense, to take only the closed box or to take both. My decision reasonably affects for me the probability that the closed box contains $10,000 rather than nothing. How could it reasonably fail to do so? For any onlooker, my decision *obviously* affects that probability: it has not failed in a single one of the preceding 2000 cases. And I am, in this matter, in exactly the same position as the onlooker. The onlooker may not know how to explain it; I may not know how to explain it: but, if I do not dismiss the psychologist's past predictions as coincidental, my own choice must affect my estimate of the probability that one situation obtains or that the other does. Moreover, my choice of strategy is the *only* circumstance of which I know that bears on that probability. I thus have a choice between doing something that will, with a very high probability, result in my getting $1000 and doing something that will, with a very high probability, result in my getting $10,000. Plainly, Bar-Hillel and Margalit were right to conclude that the rational thing for me to do is the second. *After* I have done it, the rules governing the assertion of counterfactual conditionals may entitle me to assert, 'If I had taken both boxes, I should have got $11,000'; but that is only a remark about our use of counterfactual conditionals. *Before* I make my choice, I should be a fool to disregard the high probability of the statement, 'If I take both boxes, I shall get only $1000'. That is not merely a remark about our use of the word 'probability', nor even about our use of the word 'rational', but about what it is rational to do.

FURTHER READING

Bar-Hillel, M. and Margalit, A., 'Newcomb's Paradox revisited', *British Journal for the Philosophy of Science*, 23 (1972), pp. 295—303.

Cargile, J., 'Newcomb's Paradox', *British Journal for the Philosophy of Science*, 26 (1975), pp. 234—9.

Dummett, Michael, 'Bringing about the past', *Philosophical Review*, 73 (1964), pp. 338—59; reprinted in *Truth and Other Enigmas*, see below.
In reply to the above: Gorovitz, Samuel, 'Leaving the past alone', *Philosophical Review*, 73 (1964), pp. 360—71.

Dummett, Michael and Flew, Antony, 'Symposium: can an effect precede its cause?', *Proceedings of the Aristotelian Society*, suppl. 28 (1954), pp. 27—62. Michael Dummett's article is reprinted in Dummett, Michael, *Truth and Other Enigmas* (Duckworth, London, 1978), pp. 319—32.

Mackie, J. L., 'Newcomb's Paradox and the direction of causation', *Canadian Journal of Philosophy*, 7 (1977), pp. 213—25.

On Newcomb's Paradox: Nozick, Robert, 'Newcomb's problem and two principles of choice', in *Essays in Honor of Carl Hempel*, ed. Nicholas Rescher (Reidel, Dordrecht, 1969), pp. 114—46.

Wells, H. G., *The Time Machine* (Everyman's Library, Dent, London 1977), chapter 1.

Glossary

Any term that appears in bold type in a glossary entry has an entry of its own. Other technical terms are italicized.

antecedent The first clause of a **conditional** — what comes immediately after the 'if'. Thus, in 'If it rains, then the match will be cancelled', 'it rains' is the antecedent. See **consequent**.

baryon Matter is believed to consist primarily of *leptons* (literally 'light particles') which include electrons, muons and neutrinos, and *baryons* (literally 'heavy particles') which include protons and neutrons. According to current theory, the leptons are truly fundamental particles, whereas all baryons are composed of *quarks*.

black body Black body radiation is radiation with a particular kind of spectrum, or distribution of amplitudes with respect to frequency, such as would be emitted by an object with a perfectly black, i.e. non-reflective, surface that was in **thermal equilibrium** with the surrounding radiation field, emitting just as much radiation as it absorbs.

black hole A region of space in which the gravitational force is so intense as to prevent light escaping — a region, in other words, in which the *escape velocity* exceeds the speed of light. The surface within which light is unable to escape is known as the *event horizon* and represents a point of no return for incoming objects, which will then be drawn inexorably towards a point in the middle, known as the *singularity*, at which, in theory, matter becomes infinitely dense. But before this point is actually reached, any object will be torn apart by tidal forces. Black holes have the curious property that, from the standpoint of outside observers, time itself slows down as the event horizon is approached, so that if an observer outside a black hole watches an astronaut going towards the event horizon, it will seem to this observer that the astronaut never actually reaches it. From the astronaut's own point of view, however, it only takes a finite time to reach the event horizon and enter the black hole; and although he will then be unable to send messages to the outside world, messages from the outside will continue to reach him (for as long as he is able to survive). See **proper time**.

collapse of the wave function According to the standard interpretation of quantum mechanics, the effect of carrying out a measurement or observation on a quantum mechanical system is to collapse the **wave function**, which in general will take the form of a **superposition** of states corresponding to the various possible results of the measurement, into just one of the states making up the superposition, i.e. that corresponding to the result of the measurement. This conception is, in a number of ways, deeply unsatisfactory, not least because quantum theory itself appears to provide no criterion as to exactly when, in the course of the measurement process, this collapse takes place. See *eigen*-**value/vector, many worlds interpretation**.

conditional A proposition of the form 'If P (then) Q', e.g. 'If it rains, then the match will be cancelled'. See **subjunctive conditional, counterfactual conditional**.

consequent The second clause of a **conditional** (usually preceded by 'then' or a comma). Thus, in 'If it rains, the match will be cancelled', 'the match will be cancelled' is the consequent. See **antecedent**.

coordinate time See **proper time**.

counterfactual conditional A **subjunctive conditional** with a false **antecedent**, e.g. 'If Carter had succeeded in rescuing the Iranian hostages, he would have been reelected'.

determinism The philosophical doctrine according to which everything that happens (except, perhaps, at the very first moment of time, if there was one) is an inevitable consequence of some previous state of affairs. Such a view has, historically, been encouraged by two sorts of consideration: first, a theological belief in the omniscience of God, from which it has been inferred that everything that happens would follow inevitably from God's prior state of knowledge; and secondly, classical mechanics, according to which the state of a *closed* physical system at any one time, together with the forces operating, jointly determine the state of the system at any other time during which it remains closed (i.e. free from outside influence). (Note that the universe as a whole is, by definition, a closed system, since it is supposed to include everything.) The doctrine of determinism is nicely summed up in a anonymous limerick:

> There once was a man who said 'Damn!
> It seems to me now that I am
> But a being that moves
> In predestinate grooves,
> Not a bus, not a bus,
> But a tram.'

See **fatalism**.

Doppler shift The phenomenon whereby the observed frequency and wavelength of waves depends on the relative motion of the source and the observer. If the source and the observer are approaching each other, the frequency is increased; if they are moving apart it is reduced. This is found in cases as diverse as the sound of an ambulance siren, which appears to fall in pitch as it passes you, and the light from distant galaxies participating in the expansion of the universe, which is shifted towards the red end of the spectrum.

eigen-**value/vector** Corresponding to everything one might measure or observe on a quantum-mechanical system (collectively known as *observables*) there is a set of possible results or outcomes of the measurement. These are known as *eigen*-values. On the conventional view, an observation or measurement precipitates a **collapse of the wave-function** into a state corresponding to the particular result obtained. This is known as an *eigen*-state of the measured observable. What is meant by the corresponding *eigen*-vector is the **wave-function**al representation of this state, which can be thought of as a vector in an abstract function space known as *Hilbert space*.

entropy A measure of the proportion of usable energy available within a physical system, which is roughly equivalent to its degree of organization; hence the equation of entropy with disorder. According to the *second law of thermodynamics*, the flow of energy within a physical system is always such as to cause the total entropy to rise over time (and hence the usable energy to fall); perpetual motion machines would violate this principle.

equivalence, principle of The principle, first postulated by Einstein, according to which a person in a sufficiently small closed chamber has no way of distinguishing a gravitational force from an acceleratory one. Thus, if the chamber is in a rocket, an astronaut cannot, by means of any experiment, tell whether the 1 g force he is experiencing is due to the rocket's being stationary on the ground, or to the fact that its engines are firing and subjecting it to a constant 1 **g** acceleration. This principle played an important role in Einstein's development of his *general theory of relativity*.

event horizon See **black hole**.

fatalism The philosophical doctrine according to which nothing we do can make any difference to what happens, and that consequently it does not matter what we do: all action is pointless. Fatalism has often been thought to follow from **determinism**. Nevertheless, most philosophical determinists are not fatalists.

geodesic Ordinary three-dimensional Euclidean space is just one example of what mathematicians call *manifolds*; space—time is another example. The concept of a geodesic is the generalization, to manifolds in general, of the concept of a straight line. So a geodesic, in space—time, is to space—time what a straight line is to ordinary space. The **world-line** of an unaccelerated object is invariably a geodesic. Technically, a geodesic is what mathematicians call an *extremal* path: what this means is that it is a line joining two points, such that any other line joining the same two points which differs from the first by a sufficiently small margin will either, depending on what kind of manifold one is dealing with, be invariably a bit longer (as it is with lines in ordinary space) or invariably a bit shorter (as it is with **timelike** lines in space—time).

graviton According to *quantum field theory* all forces in nature operate by the exchange of what are known as *virtual particles*. Electromagnetism, for example, is believed to operate by the exchange of virtual photons. Gravitons are the postulated particles corresponding to the gravitational force.

half-life Given any radioactive particle, its half-life is the time interval by the end of which there is a 50:50 chance that it will have decayed. If it has not decayed at the end of this interval, the chances remain 50:50 that it will have decayed by the time the same interval has elapsed again. What this means in the case of a (pure) radioactive substance such as uranium is that, on average, half the substance will have decayed at the end of the interval corresponding to the half-life of the atoms of which it is composed. Radioactive decay illustrates the irreducibly statistical or probabilistic character of modern physics, in defiance of Einstein's conviction that 'the Lord does not play dice'.

Hawking radiation The phenomenon, first predicted by Stephen Hawking on theoretical grounds and never so far observed, whereby **black holes** emit **black body** radiation, of an intensity that is inversely related to the surface area of the event horizon. This means that black holes will steadily (though very slowly if they are of a mass comparable to or greater than that of the sun) decay — the amount of radiation they emit rising as they get smaller, until they finally disappear in an explosive burst of radiation.

Heisenberg uncertainty principle See **uncertainty principle**.

hidden variable theories Hidden variable theories attempt to interpret or modify quantum mechanics in such a way as to make the outcome of a measurement on a quantum-mechanical system stand in a deterministic relationship to the state of the system and the state of the measuring apparatus at the time of measurement. This contrasts with the orthodox interpretation, according to which, in general, one can only make probabilistic predictions about the outcome of measurement, given knowledge of the current state of the system. According to such theories the description of a quantum mechanical state in terms of the **wave function** is essentially incomplete; the hidden variables are supposed to embody the further specification that is needed to make it into a complete description.

hyperbola Take two straight lines, and rotate one around the other, keeping the angle between them fixed. This defines a surface of rotation which mathematicians call a *right circular cone*, but which lay people would regard as two cones tip to tip. Now consider a plane surface that slices through both cones. The curve marking the places where the plane surface meets the surfaces of the two cones, which accordingly has two parts, is known as a hyperbola. Algebraically, a hyperbola is the graph of an equation of the form:

$$\frac{x^2}{a^2} - \frac{y^2}{b^2} = 1.$$

imaginary Imaginary numbers are numbers that are multiples of the square root of minus one, i. They were introduced to provide solutions to equations that have no solutions in the form of ordinary *real* numbers, such as 45, 2/3 or pi. **Tachyons**, if they existed, would have masses such as $3.5i$ times the mass of an electron. It should be emphasized that, from a philosophical point of view, so-called imaginary numbers are no less and no more real than so-called real numbers; moreover, they both have important physical applications.

inflationary models A class of cosmological theories, of a kind first introduced by Alan Guth, according to which the universe underwent explosively rapid exponential expansion during the first few moments of its existence. This notion has been claimed to solve a number of problems that beset the traditional big-bang models, including horizon problems (see p. 45) and the problem of why, considered on any sufficiently large scale, matter and radiation appear to be distributed in such a homogeneous fashion within the observable universe. Edward Tryon has proposed a mechanism for this initial *inflation*, based on the idea that the universe arose as a fluctuation (or 'bubble') within a pre-existing vacuum. (The vacuum, in quantum field theory, is an entity which seethes with activity, in which particle—antiparticle pairs are constantly being created and undergoing mutual annihilation.)

light cone The surface defined by the space—time paths of all possible light rays through a given space—time point, P. The light cone divides the space—time continuum, relative to P, into three regions: the region outside the light cone, sometimes known as the *absolute elsewhere*, the *forward light cone* or *absolute future* and the *backward light cone* or *absolute past*. According to the orthodox view of the matter, events occurring at P can be causally affected only by events in the absolute past, and can affect only events in the absolute future. The existence of **tachyons**, however, would appear to conflict with this conventional wisdom; and it is arguable that an instantaneous, physical **collapse of the wave function** would do so too.

Mach's principle According to Mach's principle the difference between accelerated and unaccelerated motion can be defined by considering the motion of a given object relative to all the other objects in the universe, and their respective masses. Thus, according to Mach's principle, the reason why one's coffee is subject to a centrifugal force when one stirs it, and the surrounding universe (which is, after all, rotating relative to the coffee) does not, is that the mass of the coffee is exceedingly small compared to the mass of the rest of the universe.

many worlds interpretation An interpretation of quantum mechanics according to which there is no **collapse of the wave function** when a measurement or observation is carried out on a quantum mechanical system. According to the many worlds theory, what happens is that the observer herself goes into a **superposition** of states corresponding to all the different results she might have got. But these different states are mutually inaccessible; so observers always think that they have got one particular result to the exclusion of all the others, since they are unaware of the parallel universes in which they got different results. If the many worlds view is correct, **Schrödinger's cat** really is in a superposition of live and dead states, and remains so even when the closed chamber is opened and an observation of the cat is made.

Newtonian physics or mechanics The system of mechanics instituted by Newton and defined by his three *laws of motion*. To be contrasted with relativistic and quantum mechanics.

phase space Phase space is an abstract space, different points of which correspond to different possible states of some given physical system. Thus the instantaneous state of a single point-particle constrained to move in a straight line could be represented by a two-dimensional phase space in which one dimension represented the particle's momentum and the other its position on the line. The development of the system over time would then define a path through the phase space.

proper time In the theory of relativity, proper time, for a given object, is time as it would be measured by an ideal clock that moves with the object in question. The elapsed proper time between two points on an object's **world-line** is in fact a measure of the length, in space—time, of the corresponding segment of the world-line. (Clocks, in other words, are in the business of calibrating their own world-lines.) *Coordinate time*, by contrast, is time as measured by an ideal clock which, relative to one's choice of a coordinate frame, is at rest at the origin. Thus, if we take the earth as our frame of reference, then an astronaut who flies towards a **black hole** will take an infinite coordinate time to reach it, but only a finite proper time. In astronomy, proper time is time as measured by a *representative particle* — one, that is to say, which is a typical participant in the overall expansion (or subsequent contraction) of the universe.

reductionist(ic) Said of any philosophical theory according to which something, X, consists of nothing over and above something else Y, i.e. which *reduces* X to Y. Leibniz' view of space, according to which it consists of nothing over and above the spatial relations that hold between objects, is an example of a reductionist theory. (Leibniz was also, incidentally, a reductionist about relations, since he claimed that they could be reduced to the intrinsic attributes of the things related; but that is another story.)

Schrödinger's cat, paradox of See **superposition**.

spacelike In the theory of relativity, a spacelike line is a path in space—time every two points on which are, relative to any frame of reference one cares to choose, separated by a spatial distance which *dominates* the temporal interval between them, in the sense that if they are, say, spatially separated by one light year, then they will be temporally separated by less than a year. See **timelike**.

subjunctive conditional A **conditional** which states what *would* or *would have* happened if something else *were to* happen or *had* happened, e.g., 'If I were to drop this vase, it would break'. These differ from *indicative* conditionals, such as 'If John is not at work, then he is at home' in that they can be false even when the **antecedent** is false. Thus if John actually is at work, the above conditional cannot be false; but the subjunctive conditional 'If John were not at work, then he would be at home' might be. Perhaps the truth is that, if John were not at work, then he would be out fishing, since that is where he spends all his daytime leisure hours. Precisely how subjunctive conditionals are to be assessed for truth or falsity is a matter of lively debate amongst philosophers.

superposition Given a set of physical states in quantum mechanics, it is possible to multiply each state, or strictly its mathematical representation, by any number one likes, add the various products together, and get a new possible physical state, which is then known as a superposition of the states one started with. This is very odd, because the sum of an electron's being, for example, in place *a* and its being in place *b* has no obvious commonsense meaning. But from such a superposition one can read off the respective probabilities that, if one measures its position, one will find it in one place or the other. The paradox of **Schrödinger's cat** purports to show that if microscopic objects can be in superpositions of states, so can macroscopic objects. The formalism of quantum mechanics seems to imply that if the fate of a cat in a closed chamber is made to depend, for example, on whether at least one atom in a radioactive substance decays within a given time interval, then the cat itself will go into a superposition of alive and dead states! See **many worlds interpretation**.

tachyon Tachyons are postulated faster-than-light particles. These are not logically ruled out by the theory of relativity, since the latter says only that it is impossible for a particle to cross the speed of light barrier. Tachyons, which would have to have **imaginary** mass, could not be decelerated to less than the speed of light, just as ordinary particles cannot be accelerated to beyond the speed of light. Technically, tachyons could be defined as particles with **spacelike world-lines**.

thermal equilibrium A physical system is in a state of thermal equilibrium when the temperature is uniform throughout. This means that, as far as kinetic energy and radiation are concerned, the entropy of the system is at its maximum possible value. See **black body**.

timelike A timelike line is a path in space—time every two points on which are separated by a temporal interval which dominates the spatial distance (from the point of view of every frame of reference). See **spacelike**.

two slit experiment When light is passed through a screen with two slits close together, and allowed to fall on a second, image, screen, one gets *interference fringes*. This was cited by Thomas Young, who devised this experiment, as evidence of the wave nature of light. It is now known, however, that one can obtain interference fringes with electrons (or indeed, in principle, with any particle). A curious feature of the experiment (a feature predicted by quantum mechanics) is that the fringes disappear, giving way to two discrete lines of light, or regions of impact, if one has a detector capable of registering which slit a photon or electron goes through: the photons or electrons then behave like particles rather than like waves — as they do when only one slit is open. Moreover, in the absence of such a detector, one still gets interference fringes, even if the photons or electrons are sent through one at a time, no matter at how great an interval. So what appear to be 'interfering' with each other are not the particles themselves, but rather their two possible trajectories.

uncertainty principle The principle, first put forward by Werner Heisenberg, according to which the more the position of a particle is constrained, the less certain it is what value will be yielded by a measurement of its position, and vice versa. Note that the uncertainty here is not uncertainty as to where the particle really *is*, or what its actual momentum *is*; for according to quantum mechanics, particles do not possess precise positions or momenta. Rather it is an uncertainty about what outcome would result from a measurement or observation of its position or momentum.

wave function A mathematical function that is used to represent the state of a quantum mechanical system. If the quantum mechanical state is represented by a wave function that is a function of position, then the absolute value of the square of the function, for a given value of the position variable, is proportional to the probability that, upon measurement, the particle will be found in the immediate neighbourhood of the corresponding position.

wave—particle duality The principle according to which particles, such as photons or electrons, can behave in different contexts either like waves or like particles. See **two slit** experiment.

world-line At any given time, an object will (waiving quantum-mechanical considerations) be in a particular place. From the standpoint of the theory of relativity, this (if we ignore the fact that it is spatially extended) amounts to its occupying a particular space—time point. The world-line of an object is then the line comprised of all the space—time points that an object occupies in the course of its existence. Ordinary objects, as distinct from **tachyons**, have **timelike** world-lines.

Bibliography

GENERAL INTRODUCTIONS (NON-TECHNICAL)

Shallis, Michael, *On Time* (Penguin, Harmondsworth, 1983).
Whitrow, G. J., *What is Time?* (Thames and Hudson, London, 1972).

PHILOSOPHICAL ANTHOLOGIES

Smart, J. J. C., *Problems of Space and Time* (Macmillan, New York, 1964).
Gale, Richard M., *The Philosophy of Time* (Macmillan, London, 1968).
Freeman, E. and Sellars, W., *The Philosophy of Time* (Open Court, La Salle, 1971).

PHILOSOPHICAL WORKS SPECIFICALLY ON TIME OR SPACE

Lucas, John, *A Treatise on Time and Space* (Methuen, London, 1973).
Van Fraasen, Bas C., *An Introduction to the Philosophy of Time and Space* (Random House, New York, 1970).
Swinburne, R. G., *Space and Time* (Macmillan, London, 1969).

MORE TECHNICAL WORKS

Grunbaum, A., *Philosophical Problems of Space and Time* (Reidel, Dordrecht, 1973).
Newton-Smith, W. H., *The Structure of Time* (Routledge and Kegan Paul, London, 1980).
Reichenbach, H., *The Philosophy of Space and Time* (Dover, New York, 1958).

Reichenbach, H., *The Direction of Time* (University of California Press, Berkeley, 1972).

Nerlich, Graham, *The Shape of Space* (Cambridge University Press, Cambridge, 1976).

Sklar, L., *Space, Time and Spacetime* (University of California Press, Berkeley, 1974).

PHILOSOPHICAL WORKS WITH USEFUL CHAPTERS ON
TIME OR SPACE

Smart, J. J. C., *Between Science and Philosophy* (Random House, New York, 1968).

Smart, J. J. C., *Philosophy and Scientific Realism* (Routledge and Kegan Paul, London, 1963).

Lucas, John, *Space, Time and Causality* (Oxford University Press, Oxford, 1984).

ZENO'S PARADOXES

Grunbaum, A., *Zeno's Paradoxes in Modern Science* (Weslyan University Press, Middletown, 1967).

Barnes, Jonathan, *The Presocratics* (Routledge and Kegan Paul, London, 1979).

SCIENTIFIC ANTHOLOGY (RATHER TECHNICAL IN PARTS)

Landsberg, P. T., *The Enigma of Time* (Adam Hilger, Bristol, 1982).

ELEMENTARY INTRODUCTIONS TO RELATIVITY

Coleman, James, *Relativity for the Layman* (Penguin, Harmondsworth, 1969).

Eddington, Arthur, *Space, Time and Gravitation* (Cambridge University Press, Cambridge, 1920).

Einstein, Albert, *Relativity, the Special and General Theory: A Popular Exposition* (Methuen, London, 1920).

Epstein, Lewis C., *Relativity Visualized* (Insight Press, San Francisco, 1983).

SPECIFICALLY ON 'CLOCK PARADOX'

Marder, L., *Time and the Space Traveller* (George, Allen and Unwin, London, 1971).

POPULAR SCIENTIFIC WORKS WITH USEFUL CHAPTERS
ON TIME OR SPACE

Gamow, G., *One Two Three . . . Infinity* (Bantam, New York, 1971).
Narlikar, Jayant, *The Structure of the Universe* (Oxford University Press, Oxford, 1977).
Sciama, D. W., *The Unity of the Universe* (Faber and Faber, London, 1959).
Layzer, David, 'The asymmetry of time', *Scientific American*, Dec. 1975.

MORE TECHNICAL SCIENTIFIC WORKS

Gray, Jeremy, *Ideas of Space* (Clarendon Press, Oxford, 1979): more historical than philosophical.
Davies, P. C. W., *The Physics of Time Asymmetry* (Surrey University Press, Guildford, 1974).

WORKS OF SCIENCE FICTION

Wells, H. G., *The Time Machine* (Everyman's Library: Dent, London, 1977).
Wyndham, John, *The Seeds of Time* (Penguin, Harmondsworth, 1969).
Benford, Gregory, *Timescape* (Pocket Books, New York, 1981).

Index